Emancipatory Urbanization

Emancipatory Urbanization

On the Independence of Mountain Territories in relation to

Mega – City clusters:

A Transect Approach

by

Dan Narita.

Bibliographical information held by the German National Library
The German National Library has listed this book in the Deutsche Nationalbibliografie (German national bibliography); detailed bibliographic information is available online at http://dnb.d-nb.de.
1st edition - Göttingen: Cuvillier, 2021

Nonnenstieg 8, 37075 Göttingen, Germany
Telephone: +49 (0)551-54724-0
Telefax: +49 (0)551-54724-21
www.cuvillier.de

1st edition, 2021
This publication is printed on acid-free paper.

ISBN 978-3-7369-7517-0
eISBN 978-3-7369-6517-1

Biography:

Dan Narita is an architect, urbanist and researcher based in Hong Kong. He holds a Ph.D. in Urbanism and Regional Planning from the Università Iuav di Venezia and conducted research under Prof. Paola Vigano at the École Polytechnique Fédérale de Lausanne. He joined Prof. Vigano's Research group focusing on Urban Design, Territories & the Horizontal Metropolis and was co-supervised by Prof. David Grahame Shane at Columbia University, New York. With an interest in Research by Design, he balances academic and practice-based knowledge by collaboration with international architecture & urban design practices from Beijing, Hong Kong, London, Paris and Vienna since the past 18 years. Prior to specializing in sustainable urbanization in the Asian context and in particular China, he has completed a MSc. in Advanced Environmental and Energy Studies at the Centre for Alternative Technology – Graduate School of the Environment in the UK, and a Diploma in Architecture at the Architectural Association in London. He is a lifetime member of The International Society of City and Regional Planners (ISOCARP). As part of his current research commitments, he studies the potential of rural territories as cities and their relationship to mega-city regions. His research interest in urbanism include; nature-based urban restoration, emerging concepts of urbanity, ecological regeneration & resilience, multi-functional landscapes and sustainable development.

Preface

Discussing the development of remote mountain communities requires interdisciplinary thinking and acknowledges the interrelatedness of different fields of knowledge. Coming from the perspective of an architect and urbanist, in this book a project-based and propositional approach was chosen to engage with social and environmental issues in inhabited territories. The work is organized in five main parts. The Research by Design studies is considered the main body of this research. As part of the baseline research for the design charrettes, extensive on the topic literature review, two dimensional cartographic analysis, fieldwork interviews and analytical essays have been generated.

As the title of the work indicates, a transect approach as a methodology for regional design is emphasized. There are two key reasons for opting for a design-led approach to the rural-urban discourse in China: Firstly, during conducting the baseline research, it became evident that the National New Type Urbanization Plan 2014 – 2020 (NTUP) promulgated for a five-year period, is too short for transforming urban development in the PRC. Hence, discussions beyond the year 2020 were needed. Furthermore, there is hardly any research work which tries to construct design scenarios for how to implement the objectives of China's NTUP. Some of the Alibaba Poverty Alleviation projects in rural areas suggest that e-commerce delivery platforms could lift communities out of poverty (see Fig.1.01). This book explores alternative approaches to the notion that online retailing ecologies could solve social and environmental problems in rural communities.

Secondly, a transect approach for regional design studies in the Chinese context is still relatively new. In Western rural-urban planning the transect approach has a long-standing tradition. The highly intertwined rural and urban uses of in China are a challenging but also exciting opportunity to test ways in which regional planning in transects can be applied for the rural-urban context in China.

The design investigations and propositions are informed by the analytical and fieldwork stage of the research work. Alternative ways of engaging with and addressing sustainable development issues in disadvantaged rural territories are derived from proposals in design studies at four regional scales. The four main scales of investigation are the Pearl River Delta scale in Guangdong Province, Dongjiang River Basin including the Hakka Mei County (See Fig.1.02), and the Mountain valley scales, and community level scale of local people.

Fig. 1.01: The Chinese newspapers China Daily proclaims the rebirth of rural areas in China as E-commerce platforms offer opportunities to open new distribution channels for selling goods. The "Taobao Villages "are still a relatively new concept and it is too early to understand the positive and negative externalities, to assess whether E-commerce platforms are a solution for poverty alleviation in rural areas. Source: Song Chen, ChinaDaily, 2018. Retrieved 05.07.2019, https://www.chinadaily.com.cn/a/201801/27/WS5a6baf51a3106e7dcc137153.html

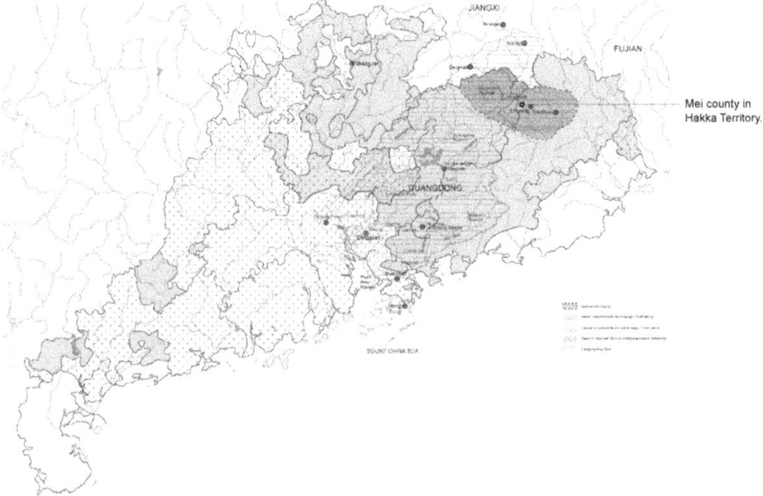

Fig. 1.02: The map of Guangdong province shows Mei County in the upstream zone of the Dongjiang River hydrological system. Additionally, the light brown area shows extent of the Hakka ethnic distribution in Guangdong. It is necessary highlight that definitive boundaries for ethnic cultural regions and hydrological systems cannot be determined. For instance, the transitions between the Hakka and Canton cultural zones are often fluid or overlapping. Source: Map elaborated by the author.

The literature review discusses rural-urban transformation in relation to urbanization in China and the potential of mountains as alternative settlement spaces. Patrick Geddes's Cities in Evolution and Terry McGee's Desakota concept underpin this inquiry. China specific knowledge such as the reforms to the Hukou household registration system, and the Communist Party of China's (CPC) National New Type Urbanization Plan 2014-2020 are introduced. Perspectives from different researchers studying development issues in China such as Hou's notion of Community Capitalism in China, as well as work on 'Taobao Villages' are presented. The literature review has helped to define the hypothesis and research questions for this study.

Fieldwork excursions to the Dongjiang River Basin, Fengshuba Reservoir, Xingning, Mei County, Yannanfei Tea Plantations, including several visits to remote Mountain Hakka Village communities were integral for the mapping and data collection stage in the territories. The site visits also gave opportunities to conduct formal and informal interviews with the local residents providing primary sources of in-formation.

In summary, the book comprises of the following five main parts:

Part One, which includes the literature review presents the problem field, the main hypothesis and research questions laying out the research objectives. This part introduces the idea of mountains as an underestimated resource and backbone territory for metropolitan areas.

Part Two provides a macro-scale overview of the urbanization process in China and its system of mega - city corridors (see Fig. 1.07). It illustrates how this network of urbanization corridors is intended as a framework for the development of landlocked territories in the PRC. However, despite the country-wide development strategy such as the Road and Belt Initiative, remote territories remain underdeveloped. This part is devoted to the contextualization of the research area in the Dongjiang River Basin with its Hakka Territory in Guangdong. The subsection on the Opportunities and Challenges of the Dongjiang River Basin & the surrounding Hakka Territory introduces the dependencies between the study areas in Mei County and the Pearl River Delta metropolitan agglomeration.

Part Three is dedicated to the notion of the design project as a knowledge producer. Scenario constructions at the Dongjiang River Basin scale, at the mountain valley scale and grassroots level scale of five individual local characters have been studied. The knowledge and insights for the scenario constructions have been gathered through interviews with local inhabitants. Alternative livelihood scenarios are proposed. The synthesis of potential environmental and welfare-oriented interventions in mountain valley communities and street transformations in the settlements are then presented at the end of this part. The objective of the design interventions is to enable improved livelihood independence for communities in upstream mountain regions. In summary, a potential repertoire of possible long-term interventions for 2050 and beyond for disadvantaged territories is exemplified. The propositions are not intended as prescriptive or absolute conclusive scenarios. There is neither a particular sequence nor a set of combined interventions suggested as the only problem-solving answer. Instead of giving a definitive resolution to social and environmental problems, the idea of having repertoires of possible interventions prepared for communities to maintain economic and resource independence is the culminating recommendation.

Part Four is dedicated to the Transect as a planning method. Some of the influential planners, designers and scientists who used the transect method for their research or design projects are presented. An overview of the concept of regional design is provided. The necessity to work at the regional scale to address global concerns including urbanization, climate change, food insecurity, migration and environmental pollution is highlighted. Parallels to post-WWII Europe are drawn when 'Géographie Volontaire' as a field of knowledge emerged out of the need for large-scale national reconstruction projects. Furthermore, the transect as a method for the analysis and design of territories is treated in this part. Additionally, contemporary large-scale urban ecology reconstruction and planning projects conceived during the New Type Urbanization Plan 2014 – 2020 in China are presented. The projects shown are influenced by Ecological Urbanism concepts, the Garden City movement, and the idea of rural-urban hybrid territories.

Part Five is the concluding part, discussing the reassessment of the transect method for regional design projects. The current popularized approaches to rural development by online retailing corporations in China are compared to planning questions addressed by the Patrick Geddes and New Urbanist Movement transect methods. Regional development issues such as the 'relationship of people in their geographic context' and the 'rural-urban continuity' as themes of transect methods underscore the relevance of transect planning equally pertinent for the Chinese context. Moreover, the rethinking of the transect method is outlined by suggesting combining some of the key features from different 'schools of thought' into hybrid methods and application of the transect for design projects. Finally, three future research questions are outlined for further study as the next urbanization phases unfold.

Hong Kong, 24.10.2021

Dan Narita.

Table of Contents

Part Three

Research by Design Studies:

Part Four

Part Five

Part One

Mountain Territories as an Alternative Urbanity.

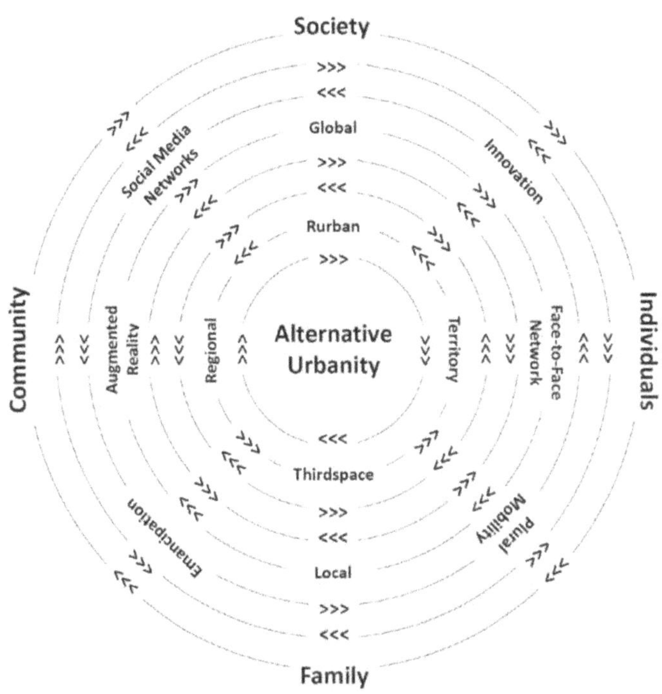

Fig. 1.03: The diagram describes the category of an 'Alternative Urbanity' in the centre of a circular diagram with subcategories relevant to this research project. Source: by the author, 2018.

1.1. Introduction:

The Research Objective:

A deeper rethinking of a people-oriented urbanity for future cities in China inevitably extends the scope of inquiry into territories which includes the countryside and mountain regions. The re-organization of regional spatial configurations is pertinent in times of social, political, economic and environmental transformations taking place. The shorter cycles, during which cities emerge and decline, reveals the brutality of the contemporary planetary urbanization (Brenner and Schmidt, 2015). The human and environmental costs of the shorter urban development cycles become more rapidly visible than in previous slower stages of urbanization. Changes in the location of population centres and shifts of the concentration of economic activities can be traced over longer durations of time. Transitions in the system of urbanization can be caused by revolutions by the citizen, climate change, environmental disasters, foreign policy & security considerations, technological innovations and migrations. In other words, changing priorities in urbanization processes can lead to the rediscovery of under -developed territories across space and time. This is not to say, that established urban centres loose their relevance or competitive advantages. It is rather, that overlooked regions can offer alternative spaces for settlement. Therefore, during periods of a regional or a national crisis, the necessity to reconstruct territories and its ecologies becomes pertinent (Cupers, 2016). Moreover, changes in the dynamics and relationships between dominant and peripheral territories become redefined. Consequently, the boundaries between what constitutes 'centre' and 'periphery' can be renegotiated.

Overtime, a slow integration of peripheral mountain areas into urban systems, in incremental steps, may establish a new order between cities in the future. The independence of mountain territories from metropolitan areas can catalyse new synergies for urbanization and alternative livelihood strategies in rural communities. The overdevelopment of the South-East coastal cities in Greater China has caused environmental degradation, unbalanced economic growth, and acute social disparities between the developed Pearl River Delta and the remote mountain regions in Guangdong Province. The Upstream Dongjiang River Basin in the Hakka Mei County of Guangdong has been chosen as a testing ground for research on rural-urban habitats located in the hinterland of the coastal zone. Motivated by an interest to think through an urban transition by design the investigations try to unfold the possibility of an alternative form of urbanity in the mountain territories. Planning for a phase beyond the National New-Type Urbanization Plan 2014 – 2020, the next phase of urbanization in China targets a 70% urbanised population by 2030 (see Fig. 1.04). The ambition for the following urbanization stages is to take into consideration emerging new livelihood ecologies and sustainable development goals. China during the advanced 'New Normal' development stage (see Fig. 1.05), a stage of economic slowdown, is in the process of adapting communities for the needs and interest of an aspiring middle-class society (see Fig. 1.12).

Context: Urbanization targets for China by 2030+

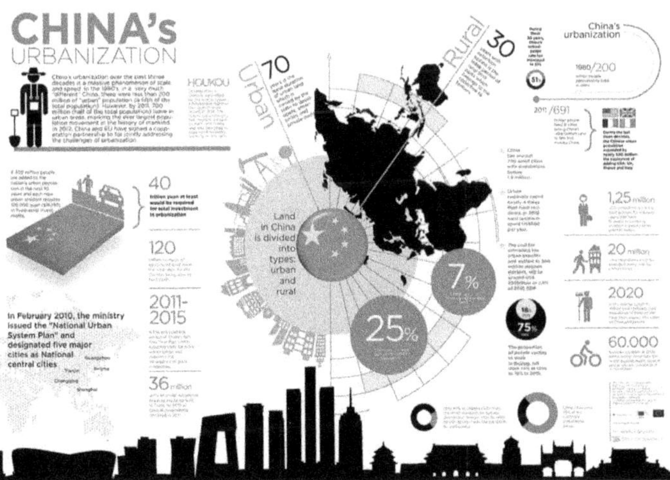

Fig. 1.04: The diagram explains the urbanization process in China since the 1980s and future tendencies. An increase of the current urbanization rate of over 50% since 2011, to an urbanised

population of 70% may be the future path by 2030. It is expected that the approximately 200 to 300 million floating population in China will settle in new smaller and medium size cities. Source: China's Urbanization, produced by DRAGONSTAR: HUKU INFOGRAPHIC. Retrieved 06.08.2019, http://www.dragon-star.eu/wp-content/uploads/2013/03/HUKU-INFOGRAPHIC.pdf

New Normal: Economic projections for an urbanization led growth strategy.

The development of smaller cities is expected to sustain future economic growth.

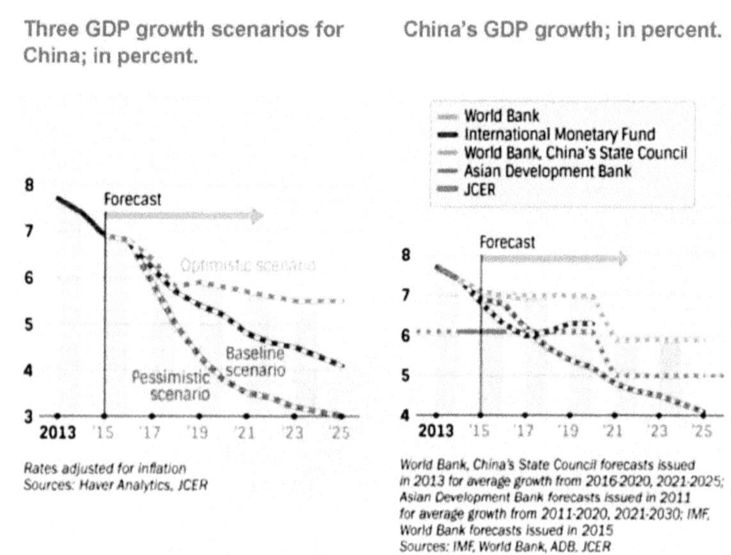

Fig. 1.05: A slower GDP growth rate referred to as the 'New Normal' will likely be experienced not only in China. The above GDP growth rate scenarios show possible trajectories of the economy, with Optimistic, Baseline and Pessimistic Scenarios. The baseline scenario indicates a rate of +/4% by 2025. Source: Nikkei Asian Review, July 2015.

From the perspective of maintaining economic growth, the increasing of the numbers of medium and small sized cities, is still regarded as an economic strategy in China (see Fig. 2.08). Furthermore, the settling of China's 'floating population' is a yet to be resolved issue (see Fig. 1.06). Equally important is maintaining the inhabitability in cities, which involves reinstating a clean 'Blue-sky', better air quality in urban areas, and the urgency for environmental remediation. Shifting the attention from metropolitan cities to settlement structures in mountain territories may imply a de-concentration of the population in established cities. The settlement of people away from the coastal cities in the not yet developed habitats requires regional design strategies sensitive to local ecologies. Dispersed, low-density settlement structures in the valleys of 'sub-alpine' territories may cater for 'urban life-style' options and environmentally-friendly production ecologies, without living in congested metropolitan areas.

Compelled by trying to search for an alternative urbanity, this research studies mountain valley settlements in Guangdong located close to the border of Jiangxi and Fujian Province. The unbalanced economic relationship between the Pearl River Delta metropolis and its mountainous hinterland is studied in a series of Research by Design explorations.

Cartographic studies look at possibilities to support independence of mountain settlements in the region. Attention is paid to basic farm land preservation, protection of the hydrological network as a shared environmental capital, and the creation of habitable settlements outside of urban areas. The global challenges of food, water and migration trigger interrelated new urban questions. The design studies construct potential scenarios considering both bottom up and top down design interventions. Respecting exiting social and new transactional networks in the territory, the design scenarios look at how to blend-in new linkages, livelihood ecologies enhanced by novel technologies, efficient mobility systems, remediated environmental resources, and green infrastructure.

Ultimately, the scenarios outline an independent urban future for the Mei County, a Hakka ethnic territory, located at the upstream region of the Dongjiang River Basin. The rediscovery of mountain territories as a habitat and environmental resource is put forward for a new phase of urbanization in China. Emerging, liveable cities responsive to climate change, scarcity of resources, widening social inequalities and the dominance of global city networks may not be found in densely populated urban areas. The underestimated potential of mountains with dispersed settlement structures become defined as an alternative people-oriented urbanity. The superimposition of old and new transactional networks as hybrid multi-functional spaces deliver modernity and resilience for new type mountain cities.

Mountains as a Resource:

In mountain research, as a field of knowledge, the relevance of mountains as a productive ecological resource, essential for the eco-system stability of larger regional configurations cannot be overlooked. The metropolis as a dominant urban category in the study of cities overshadows mountains as settlement spaces for evolved and contemporary communities. As the environmental carrying capacities in metropolitan agglomerations continue to be exhausted, paying more attention to mountain regions as part of urban systems becomes more relevant than ever. There are two important reasons for exploring the transformations in mountains communities taking place. The transformations are linked to behavioural changes of people. Firstly, the adoption of 'urbanised' behaviour of people living in the countryside and mountain areas is increasing. Partly, this phenomenon is occurring due to the migration of family members, and information exchanges between urban and rural communities. Secondly, given the behavioural changes of how people live, work and consume resources in rural areas, it raises questions about sustainable development strategies to avoid the negative impacts of a new generation consumerist society and shorter economic cycles. Given the environmental challenges in Chinese cities, questions about how more equitable settlements can be enabled is driving this research. An urgent reason for the search for urbanity in outside of densely populated areas is the injustice of spatially compromised, and socially underprivileged segments of the population in large cities in China.

The negative externalities of market-oriented real estate developments such as land-grabs, the diminishing of arable fertile land, and the widening gap between rich and poor people are serious issues undermining society and future generations. It is also critical to reflect on how the retrofitting of the Chinese cities may be undertaken by planners and architects in 50 years' time. Will it be possible to remediate destabilized eco-systems and socially fragmented communities in cities? Will people move away from overcrowded urban areas to low density settlements in the countryside? How could such a process be further supported through development policies? New profit-driven urban development continues to engulf traditional Chinese villages, and a mosaic of mixed farming and built-up urban-rural areas are the outcome of uncoordinated large-scale development projects. Over-population, issues of 'illegal' residents in cities and the environmental pollution are the unresolved challenges requiring changes of our development pattern for sustainable urban futures.

The settlement of China's Floating Population in cities and Hukou Reforms is one of the challenges yet to be addressed in future urbanization phases in the PRC.

Fig. 1.06: Image shows Chinese migrant workers from rural areas seeking employment opportunities in cities in Greater China and South East Asian countries. Source: Aris Chan, Feb 2011, China Labour Bulletin, TWC2.

Moving beyond the binaries of 'urban' and 'rural' territories:

Hybrid descriptions for territories as 'rurban' indicate the gradual transformations taking place in cities and the countryside. Increasingly, the simplifying categories of 'urban' and 'rural' territories no longer adequately describe current urbanization processes. In particular, in the Asian and Chinese context the transformations of urbanised and rural regions manifest the emergence of novel types of polarities between the city and the countryside. The growth of mega-city regions causes concern over the longevity and resilience of urban agglomerations. The countryside, on the other hand, is steadily catering for similar life-style choices as in cities.

The challenges for maintaining resilience in cities include climate change, flood risk, social inequality, food insecurity, economic stagnation and the migration of people. The development of Chinese mega-cities was partly supported by the supply of environmental resources, land and surplus labour from the peripheral and agricultural regions. If the growth of mega-cities in China has reached a point of unsustainable overdevelopment, then vulnerabilities of urban areas inducing a potential collapse of the mega-city (Shane, 2017) requires preparation for a crisis. In a possible decline of the mega-city, remote rural areas may develop into a 'back-up' resource for larger metropolitan areas.

For instance, the remoteness of mountain regions could be turned into a competitive advantage. Hence, increasing the independence of such regions as an alternative low-density city, while minimizing its dependence on the overpopulated urban agglomerations. Mountain regions, protected by environmental policies and fairly negotiated agreements, can offer environmental capital such as water, the capturing of pollution, land for food production, sites for recreation and accommodation of populations. Such benefits could lessen the burden imposed on metropolitan areas. The transformation of regional interdependencies including ecological, community and information based linkages binding together the mega-city and the hinterland will be critical for finding more reciprocal urban-to-rural relationships.

Challenging the idea of high-density cities as the only path towards sustainability, the objective is to further extend the idea of multifunctional uses of land in the mountains, to include new networks of livelihood ecologies. Examples are illustrated in Chapter 3.4 and 3.5. Particular emphasis is placed on securing the inhabitability of settlements, social welfare provisions, preservation of environmental capital, the potential of new economic opportunities and linkages between the urban settlements and the mountains communities. The aim is to evolve a novel form of urbanity, which accommodates both urban and rural ways of living.

Mega-City Network

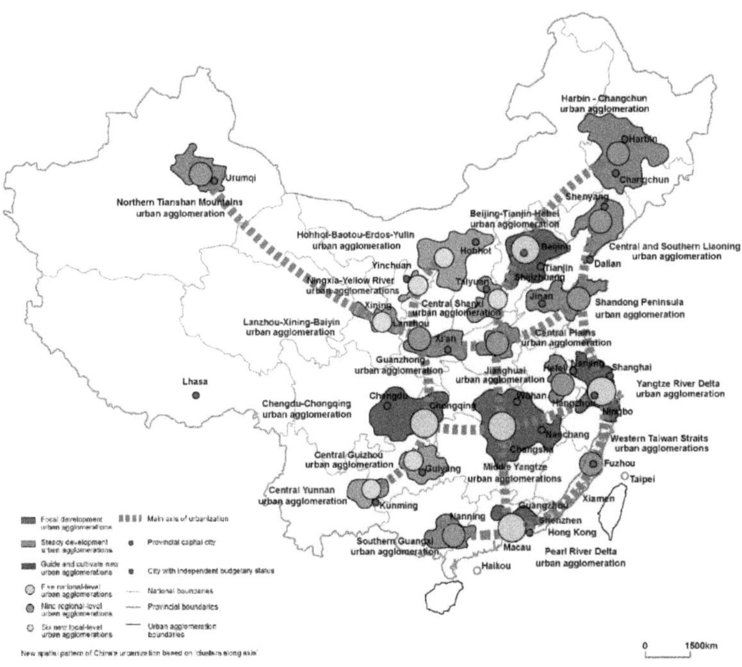

New spatial pattern of China's urbanization based on 'clusters along axis'

Fig. 1.07: The New Type of Urbanization phase has an underlying framework of 20 urban agglomerations. The mega-city urbanization system is organized in a spatial pattern of clusters along five main urbanization corridors or in other words 5+9+6 spatial structure. These are the coastal cities corridor axis, Yangtze River urbanization corridor, Eurasian land bridge urbanization corridor, Harbin-Beijing-Guangzhou urbanization corridor, and Baotou-Kunming urbanization corridor. As such they form a macro-pattern of urbanization for Greater China. Credits: Chuanglin Fang, Journal of Geographical Sciences, 2015, 25(8): pp.1003-1024.

1.2. The Conundrum:

The rise of Chinese mega-cities has been enabled by releasing environmental capital, land and labour from rural and mountainous regions. By providing the metropolitan regions with ecological resources from less affluent regions, peripheral territories have suffered environmental decay (see Fig. 1.08). Further, local communities have suffered a loss in social cohesion and disconnection from family farm land. Such social issues are the result of a remittance economy and the labour migration of peasants seeking employment opportunities in urban centres.

Escalating land and property prices associated with the ever growing metropolitan areas exacerbate the economic inequality between affluent and deprived communities. Due to exploitation of agriculture and industrialization in the rural hinterland, impoverished mountain territories in China fall into a state of crisis, struggling to prevail under the pressures of global urbanization.

While the exchanges of resources between mega-city regions and the countryside in China are principally benefitting the economies in urban areas, at the same time metropolitan agglomerations are confronted with acute environmental, social and public health issues, due to a consumption and growth oriented economy. Unsustainable development patterns in metropolitan areas have not adequately prepared for potential risks arising from global challenges, including food shortage, environmental disasters, and the urban population growth.

The lack of spare ecological capacities in cities, to cope with unanticipated, catastrophic events in the environment, and the catering for the welfare security for communities, calls for a restructuring of the relationships between the metropolis and the mountains. Instead of mountains as dependent, peripheral territories, the redefinition of mountains communities, as self-sufficient settlements is crucial for a sustainable development agenda. Ultimately, more emancipated mountain communities could reinforce better relationship with neighbouring cities at a regional scale, and strengthen synergies with mega-city clusters.

Fig. 1.08: Garbage covered in green protective nets shown as traditional Landscape Paintings reminds the viewer of nostalgia about the countryside and the issue of environmental degradation in rural areas in China. Credits: Artwork by Yao Lu.

1.3. Literature Review: On the potential of urbanizing of mountain settlements in China.

Introduction:

Bridging the longstanding rural-urban divide requires appropriate strategies to urbanise remote mountain settlements in China. The notion of an alternative urbanity in mountain territories is an urgent proposition, as it may help to reduce poverty and a diminishing population. Contemporary architects, regional planners and policy makers should embrace a close interaction approach when developing these underappreciated regions. In particular, it is important to grow both city and environment with close interaction, as per the idea by Patrick Geddes (1915). Terry McGee's (2008) idea of 'Desakota', i.e. co-existence between agricultural and urban forms of land use and settlement is also important in urban geography (see Fig. 4.02). Strategies such as expansion of urban green spaces, zero-waste policies, energy conservation, eco-district and industrial ecology plans, and municipal renewable energy guarantee that city residents enjoy a sustainable life. The emerging cities should be human- and environment-friendly. Developing rural regions calls for commitment to rural poverty alleviation as well as for emphasis on food security for the attainment of rural development. Such approaches should ensure there is alleviation of the false dichotomy or differentiation between rural and urban, as the overall aim is to bridge the gap, considering the fact that most of the future global urbanisation will take place in Africa and Asia. Bridging the gap will require a re-examination of the existing definitions of urban and rural as the activities in the two regions are increasingly linking.

Gradual hukou reforms in Chinese cities and an incremental urbanisation in mountain regions challenge the classification system of either 'rural' or 'urban' citizen. Over time, return migrants from cities could bring direct impacts through the transfer of capital and knowledge back to their villages; hence gradual transformation of the mountainous settlements may evolve into an alternative type of urbanity. Similarly, Internet accessibility and e-commerce in rural areas is promising as it opens up and improves the knowledge of rural dwellers, helping them embrace entrepreneurship, grassroots level income diversification and new livelihood strategies. This ultimately leads to economic growth and urbanisation of the rural areas, developing the so-called 'Taobao Villages' (Li, 2017). Emergence of Internet giants and e-commerce corporations i.e. BAT (Baidu-Alibaba-Ten-Cents) and JD.com in China has led to increased exchange of goods, services, market commodities, information and knowledge between rural and urban areas.

Literature Review:

Mountain Settlements in China as an Alternative Urbanity:

The concept of rural-urban development is best discussed in the text Cities in Evolution by Geddes (1915). Geddes has been trained as a biologist, and this has informed his idea that both city and environment need to grow with close interaction. According to Geddes (1915), the city is not an independent and closed organism as it is found inside a setting that takes and dissipates energy. Geddes has compared the idea of an evolving or urbanising city to that of an organism that is continuously changing. The Geddes explains the underlying meaning of two key concepts, i.e. Cacotopia and Eutopia, terms that describe the good and the bad city. Cacotopia dissipates energy and gains individual monetary gains, whereas Eutopia saves the energy that is later utilised to organise the environment, allowing enough evolution of both individual and collective life.

According to Geddes (1915), urban planning was an essential part of city evolution. In particular, urban planning offered volition to a permanently evolving organism. In urban planning, the past and present are two significant concepts that should be considered since they can be interpreted to assist in understanding or predicting the "nascent future" (Geddes, 1915). Geddes used the term 'biopolis' to imply that the city and its related planning endeavours are interrelated and are a living and open system. Similarly, Geddes (1915) interpreted evolution as a process that is fundamentally driven from within the organism. The understanding was a contrast to the existing view that external factors, i.e. in Darwin's natural selection, impacted evolution. According to the author, cooperation from the level of the cells to societies was particularly crucial since it overpowered or triumphed over the competition. In this way, cities acted as the first proof or expression of social union and evolution.

The view of evolution by Geddes utilised two different forms of organic analogy that remains necessary to understanding cities as well as to practising city planning. First, the city can be conceived as an "organic" entity regardless of whether it is interpreted as a developing or 'evolving 'depending on the conditions of its environment (Geddes, 1915). The underlying meaning of this analogy of a city as a living being suggests that town planning is an integrating theoretical as well as a practical endeavour and not merely laying down roads and built form. Therefore, city planning is not a consequence of engineering and architecture projects, but rather a discipline integrating multi-disciplinary design decisions. The second analogy regards viewing the city as an environment (built environment), and its design played a crucial role in positively influencing the contained social organism. In the second analogy, Geddes (1915) asserts that urban planners have a significant role in urbanization (or evolution). In particular, the role of a planner is to influence the beneficiary of social evolution via physical design.

THE ASSOCIATION OF THE VALLEY PLAN WITH THE VALLEY SECTION

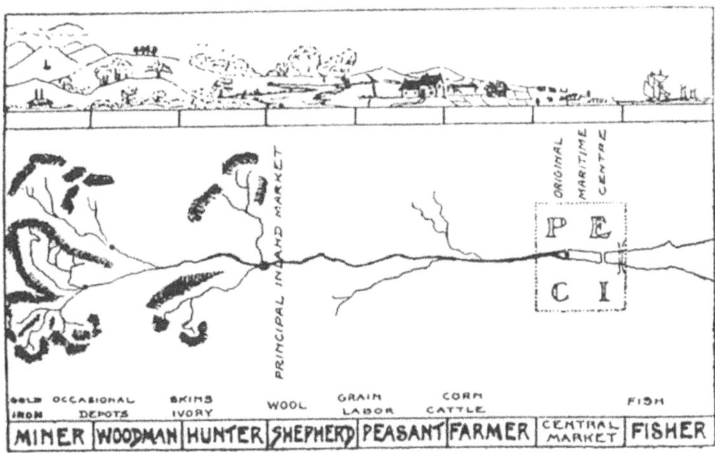

RURAL OCCUPATION & MARKET TOWN

Fig. 1.09: The Geddes Valley section: 'The Notation of Life superimposed on the Valley Region', from Victor Branford and Patrick Geddes, The Coming Polity, Published by Williams & Norgate, London,1917, pp.296.

Overall, the idea of evolutionary urbanism according to Geddes (1915) entailed part 'evolutionary' but in a non-Darwinian way, part 'developmental' that views the city as an organism, and some 'environmental' that considers the city as environment instead of an organism. The view by Geddes (1915) examines the process of urbanisation or city evolution as a combination of a different process. Such ideas remain applicable in modern-day urban and rural planning (see Fig. 1.09 & 1.18).

While Geddes' text provided relevant insights for urban planning, it not only has been out of print for more than one generation, but many planners appear to misinterpret or reinterpret his school of thought. Similarly, many leading United

States university planning programmes do not use Geddes' works as a teaching guide. The justification for the limited teaching of Geddes' works, according to many Geddes scholars, is that he did not have a theoretical framework of planning. Besides, critiques state that the work by Geddes (1915) may be obsolete in addressing power relations in modern-day cities as well as social movements.

However, despite such critique of Geddes' work, it is increasingly receiving renewed interest (Young, 2017). Contrary to the argument by many scholars, it remains evident that Geddes had a theory of planning and power, and it is relevant and applicable in contemporary issues of urbanism. The text "Free Cities and Regions"- Patrick Geddes's theory of planning by Young (2017) advances this argument. Young (2017) notes that across Geddes's voluminous output, he framed a consistent planning theory that remains valuable and applicable in modern-day environmental planning and social movements. Young (2017) does this by focusing on Geddes' theory of planning and the goals/objectives set in the theory. In his article, the specific projects Geddes produced for communities in India, Scotland, and Israel are not the main point of emphasis.

Young (2017) provides various examples supporting his position that Geddes's theory of city and regional evolution is relevant in framing and guiding metropolitan change. Geddes's theory advocates a renewed civics defined by features such as city reconstruction, pressing ethical, economic and political change, and mutual aid. A major example is a green infrastructure which is founded on Geddes's theory that emphasised on reconstructing socially and ecologically just cities (Geddes, 1915). The same concern is evident in contemporary city planning as community groups, and municipal officials in different cities continue to grapple with multiple public and environmental health issues through an expansion of urban green amenity. Information gathered from urban greening programmes increases the effectiveness of initiatives to plant trees and vegetation. Such knowledge also increases the number of informed decision-makers. Such a strategy is a perfect example of Geddes's notion of utilising geotechnic planning to improve urban evolution (Young, 2017).

In particular, integrating neotechnic technologies of satellites, aircraft and drones with computers and photography, together with citizen-based field research fosters the ideal of the regional survey. Besides, it supports a collaborative civic culture that provides an essential guide to rural-urban development (Young, 2017). In addition to this strategy, there are several other geotechnic approaches that urban planners can apply to advance contemporary efforts. As Young (2017) further posits, the civic and technological engagement that shows synergies between urban greenery provision and metropolitan populations contribute to broadening decision-making. The strategies also include zero-waste policies, municipal renewable energy, and initiatives to conserve energy, eco-district, and industrial ecology plans.

Moreover, Young (2017) supports the idea by Geddes on the pursuit of a trans-disciplinary project. Geddes combined technology, citizenship, and ecology in a new planning strategy. For effective planning of cities in contemporary societies, researchers, planning educators and practitioners should embrace the ideas by Geddes (1915), and this is through bringing them to a more active practice in course syllabi, classrooms, policy, research projects, and also citizen involvement. The above strategies can be incorporated in contemporary rural-urban planning, and equally, for the development of emerging rural settlements in China.

Another article that specifically addresses the topic of urbanisation of rural-urban territories in Asia is managing the rural-urban transformation in East Asia in the 21st century by McGee (2008). The article assesses the special features on transformation of rural-urban in East Asia in the past three decades. The emphasis is on the strategies of development that Asian governments employ. Most policies of rural-urban changes in the region disregard the ongoing commitment to alleviating rural poverty, focus on food security and also rural development. The emphasis by the East Asian governments is on the significant role of urbanisation as the major process that leads to the growth of the economy. The focus is justified by economic theories, for instance, that the economies of scale, creating and strengthening markets in cities and the increased productivity in metropolis are crucial aspects or factors that lead to development.

Nonetheless, McGee (2008) argues that this approach is ineffective as it forms a false dichotomy or differentiation between urban and rural areas. Besides, McGee (2008) informs that the approach ignores the sole role of development which is to increase connections between rural and urban regions to fully transform society instead of separating the transitions in the two regions, i.e., rural and urban. The article is particularly relevant to this book as it uses a case study of China, a perfect example due to its high population, the fact that its market socialism has unique conditions, and also the ability of the country to have an impact on the rural-urban transformation (McGee, 2008). In the article, McGee (2008) incorporates the term 'Desakota,' which he had coined around 1990. The term comes from two Indonesian words desa which means village and kota for city. The term is used in urban geography to denote spatial zones or areas on the peri-urban regions of large cities, with the major features including most intense co-existence and intensive intermingling between agricultural and urban forms of land use and settlement (McGee, 2008).The other article by McGee (2009) also discusses the concept of desakota. The article by McGee (2008) details the essence of this co-existence regarding urbanisation.

According to McGee (2008), there is the need to develop spatial sensitivity while managing the advancement in both rural and urban regions in the twenty first century. The author proposes essential strategies or approaches for active growth of the rural regions, which suit the case of impoverished mountain settlements in China. In particular, McGee (2008) argues for the replacement of the old divisions between rural and urban areas, and later embracing planning that integrates or links rural and urban endeavours for successful adoption of sustainable management strategies. The strategies to be used in developing rural regions should utilise concepts of eco-systems that link activities in urban and rural territories. In this way, there will be the preservation of ecosystems that integrate extended metropolitan region (EMR). McGee (2008) argues that successful urbanisation of rural areas should ensure industry, agriculture, and other urban endeavours co-exist in a mixed-use landscape, provide the most effective and fruitful option for preserving ecosystems within the EMR and at the same time, creating sustainable and liveable urban areas. This planned approach to attain interaction between agriculture and industrial endeavours should be in line with the existing regulations as it is founded on the land-use zoning. Such an approach is evident in the United Kingdom, where there is a green space all around the city region, and it is an essential part of the planning practice. The coexistence of rural endeavours such as agriculture and other urban endeavours like industries is what is currently called the use of "green spaces" that has become an important component of planning urban life quality in many advanced nations (McGee, 2008). The ultimate developmental progress of such co-existence would be the creation of sustainable urban areas, cities, and societies from the rural and mountain regions also in Asian countries.

McGee (2009) advances the argument on the division between rural-urban in his article The Spatiality of Urbanization: The Policy Challenges of Mega-Urban and Desakota Regions of Southeast Asia. McGee (2009) introduces the article noting that four key facts will define our thinking about urbanisation in the 21st century. The first aspect is that the future will be characterized by a majority of the world's population living in urban areas. The second fact is that most of the urbanisation in the world will take place in Africa and Asia, two continents with two-thirds of the global population, and within which at present, the proportion of people living in urban regions in most countries is below 60 per cent. McGee (2009) also notes that the new demographic realities will require a rethinking of the embedded concept of rural and urban. The fourth fact pertains the rising global economic integration, termed as globalisation, and it is creating increased chances for economic volatility and challenging the sustainability of current urban forms (McGee, 2009).The issue of changing forms of urban and rural is the most important since it acknowledges the essence of integrating or co-existing activities in rural and city regions. According to McGee (2009), the changing nature of urban and rural is significant as the concept of the rural-urban drift is particularly entrenched in the political, institutional, and social understanding of most nations in the contemporary world. Since the location of rural and urban endeavours is changing, there needs to be rethinking of the rural and urban. The author acknowledges that the non-agricultural activities, which are associated with the urban areas, are increasingly spreading into areas traditionally defined as rural (McGee, 2009).

For a sufficient understanding of the concept of rural-urban planning, it is essential to redefine the spatial patterns of the modern-day processes of urbanisation. In particular, a re-examination of current definitions of urban and rural is necessary. According to McGee (2009), the regions that were previously defined as rural, but most activities are urban, need to be reclassified. Since the coining of the term "desakota," there has been considerable debate in Asia, with neo-liberal critics arguing that such desakota region is a transitory concept that will disappear as the areas develop (McGee, 2008; McGee, 2009). In this way, the co-existence or interaction of agriculture and non-agriculture endeavours represents ineffective land use and hinders the development of successful land markets and rational space use for the creation of increased profits (McGee, 2009).

Nonetheless, McGee (2009) notes that desakota zones are particularly important for the future growth of these regions and also play a crucial role in future urbanisation. Moreover, the author opposes the standard interpretation of "globalisation" as the fundamental determinant in the growth of urban areas. McGee (2009) suggests that local endeavours at the national, regional, and sub-regional scale are particularly important in forming urban areas. The article also informs that mega-urban regions are vulnerable to global trends such as rise in food prices, cost of energy and fossil fuel and also the problem of global warming. Therefore, the author proposes that governments should develop locally derived strategies for adaptation that will be necessary for creating resilient urban areas in the 21st century.

There is a need to evaluate the kind of urban places that should be developed to satisfy liveability and sustainability. This is precisely what McGee (2010) does in his other article on Building liveable cities in Asia in the twenty-first-century research and policy for the urban future of Asia. Specifically, McGee (2010) poses a key question on the type of urban cities that should be created in Asia to fulfill the needs of liveability and sustainability over the next fifty years. The author acknowledges that Asia has a high population of almost 60 per cent of the global population, and it is also involved in rapid transition that will further add over one billion individuals to urban regions over the next five decades (McGee, 2010). For this reason, it remains vital that urban planning policies should aim at creating urban areas that are both sustainable and liveable.

There are two different visions regarding the types of urban centres that should prevail. First is the developmentalist vision that international organisations such as the World Bank often promote, and it is concerned with bettering the material conditions of urban regions. This is perceived as an essential requirement to attaining cities that are efficient "engines of growth" (McGee, 2010).The other driving force is the imperative of globalisation that compels states to create more competitive cities. The second vision seen in highly urbanised nations such as Japan, Europe, and North America holds that the major challenges to planning policy is building cities that prioritise on liveability. The vision is driven by various trends such as the understanding that problems like climate change arise, in part, due to cities' developmentalist vision. In this way, policies should focus on making cities greener through reduction of carbon footprints, promoting clean water and air, increasing green space and public parks and promoting user-friendly systems of public transport (McGee, 2010). Other endeavours include developing settlements which allow for cycling and walking paths and encouraging food production at the local community level. Young (2017) supported similar ideas of green spaces in urban regions, and this equates to co-existing rural activities of agriculture in city regions. The second trend pertains demography, and it relates to the rapid ageing of city dwellers. The third trend is in the ideas of "new urbanism" that seeks to return to participation in decision making to individuals supporting the ideas of neighbourhood, culture, the essence of heritage and conservation of the natural environment (McGee, 2010).The existing argument is that the second vision is not realistic for many Asian cities, mainly due to their rapid growth rate and the high population.

The research by McGee (2010) notes that it is possible to introduce the "sustainability" and "developmental" visions at the same time, but this should be after gaining clarification on the mechanisms driving city trends in the 21st century. The term 'building' in the article is used in the broader context of creating civic institutions such as lifestyles and practices of consumption, work systems, and cultures that comprise the liveable and sustainable urban areas. The integration of the two visions is in line with Geddes (1915) idea that environment and city should evolve with close interaction.

Urbanising the rural areas in China can also be informed by an understanding of the working mechanism of the hukou reform. The infamous household registration system (Hukou) controls or prevents free movements across locations, i.e. from rural to urban areas and vice versa in China (Lu & Wan, 2014). The new hukou system allows migrants to change their hukou status after securing employment in urban regions for a given period, but this is only in medium and small cities. In larger cities, the migrants do not enjoy the benefits as local hukou residents do, as migrants lack a full hukou status. The article by Zhan and Andreas (2015) discusses the hukou reform, with the authors noting that it has been used to serve the goal of removing many rural dwellers from their land. The article draws in various cases where peasants were forced to relocate in the city regions.

Zhan and Andreas (2015) argue that over the past ten years, hukou reforms have been associated with changes in land policy, and the fundamental objective of the reforms was to permanently ease the population pressure in the rural areas giving room for the operations of large-scale agricultural companies. Therefore, the attention of the article by Zhan and Andreas (2015) is to portray how city government, agrarian capital, and urban-industrial capital have used hukou reform to urbanise rural areas.

Agribusiness firms in China, also known as dragonhead companies (see Fig. 1.10), have significant impact on expanding agrarian capitalism in the country. The dragonhead firms contract with a considerable number of rural households, i.e., 17.7 per cent, and this gives them the authority to control over agricultural production of a huge part, specifically 23.6 per cent of cropping areas (Zhan & Andreas, 2015).Understanding the dynamics of this control is important as it sheds more light on the migration from rural to urban regions and the urban problems that emerge. The hukou system divides the Chinese population into rural and urban residents, with each category enjoying different entitlements. For example, urban residents gain better entitlements compared to their rural counterparts. The entitlements include access to various social services in the city, i.e., employment, children's education, health care, and welfare housing. On the other side, rural residents have access to land for farming and housing.

Fig. 1.10: The image by a Chinese agricultural management company or so called 'dragonhead company' is an idealized illustration of a consolidated agricultural unit. The operational unit consists of 1000mu collective farms (1mu=666.7m2). The aim of agricultural consolidation by dragon head companies is the efficient cultivation of fields and breeding of live stock through the implementation of supporting technologies, integration of land resources, unified management, and promotion of agricultural technology innovation. In this way, the unit plants oat and tea crops and feed 1000 sheep (including 200 basic ewes, 800 lambs and fattening sheep) on more than 1,000 mu of land. This operating unit can make the farmer feed 1000 sheep, plant 150,000 kg of oat or 250,000 kg of potatoes to create a direct economic benefit of 350,000 yuan/pa. Source: San Zhu Liang Group.

The hukou system significantly influences urbanisation of rural areas, but its role has limited impact in urbanising mountain areas. The system tends to force work seeking rural citizen to find accommodation in informal settlements in urban areas. The growth of agrarian capitalism is associated with hukou reforms in the past two decades, and this has helped the Chinese government to relocate rural residents to cities. However, developing isolated mountain territories in China is unattainable following Zhan and Andreas (2015) analysis of hukou reform. Hukou reforms in urban areas, i.e., legally settling peasants in cities; and other reforms in rural areas, i.e., official reclassification of rural areas as small cities, have increased the likelihood of causing indirect effects on transforming settlements in the hinterland.

The mountain territories are characterised by steep topography that will not be suitable for the farming activities of agricultural corporations. Similarly, the mountainous regions are inappropriate for large scale farming due to the high costs of transportation of farming inputs and outputs. Additionally, migrant workers in cities, leaving behind their children and elderly on small family farms, prefer holding on to their land rights in rural areas (Zhan & Andreas, 2015). In response, there is continuous resettlement of migrant workers back into the countryside, and eventually the returning workforce is likely to have trickle-down effects on communities in remote areas. In particular, there is a transfer of information, advanced technologies and state-of-the-art knowledge from cities to the rural areas, making the mountain areas gradually transform into settlements better networked with metropolitan areas.

The overall idea by Zhan and Andreas (2015) is that hukou reforms and new synergies between agrarian and industrial capital in China could lead to urbanisation of rural regions. However, Lu and Wan (2014) put forth a contrary argument noting that the existing land and hukou systems in China cause incomplete and distorted urbanisation as it is characterized as "pro-small or anti-big" cities. In their article Urbanization and urban systems in the People's Republic of China: research findings and policy recommendations, Lu and Wan

(2014) point out that the two aspects, i.e., hukou and land systems in the country are associated with institutional constraints.

The hukou system would be beneficial if it was complete in its working mechanism. Lu and Wan (2014) argue that urbanization in the People's Republic of China is incomplete, and this has severe ramifications. The abnormalities (see Fig. 1.11) associated with the system include limited urbanisation in relation to the status of development and industrialisation, discrimination against rural-urban migrants, and many urban residents not enjoying the benefits in urban society. The discrimination favours urban residents, and has caused to high levels of inequalities in the country. It is also evident that city land expands at a faster rate compared to the growth of population. Another adverse impact of the hukou system is severe distortions in the city system where there are numerous small cities but few big cities. The fact that urbanisation is not simply about an increase in urban dwellers, but also entails spatial spread of population across various cities, which informs the position of the ramification of genuine urbanization limited by the hukou system.

Nominal Urbanization Rate of the Population: ~ 17% discrepancy.

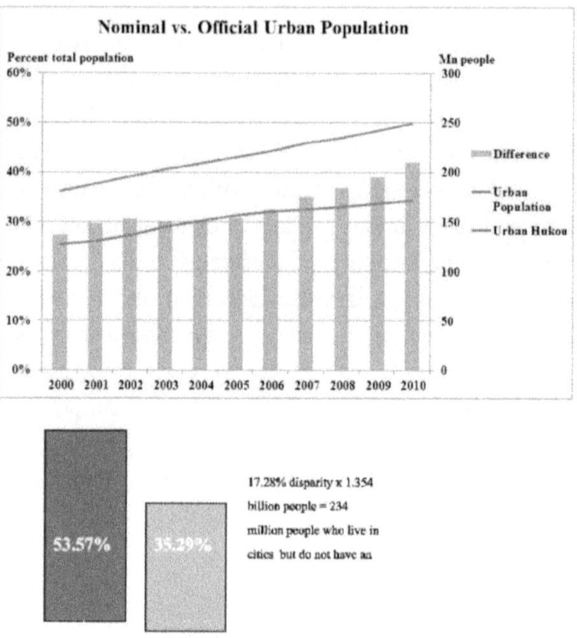

Fig. 1.11: Distortions in the Urbanization System: Depending on which sources are looked at the floating population is estimated between 120 to 274 million people in 2015. Due to the informal and temporary settlement of migrant workers an exact figure is difficult to derive. By 2020, the floating population may become up to 291 million people. Various Sources: Paulson Institute, SCMP, China Daily, China Economic Watch / Urbanization and Economic Growth China. Nicholas Borst (PIIE), Sept. 2012.

There is a need to reform the land and hukou system for effective and complete urbanisation. Zhan and Andreas (2015) supported reforming the hukou system to ensure that the benefits spread to the entire population, i.e., both rural and urban dwellers. Similarly, Lu and Wan (2014) share similar insights, giving four principles that should guide the transformation of the hukou system. The first principle is phasing out the disparities in enjoying public services as a result of the system. According to the authors, all individuals should have access to healthcare, education, social security, and public housing. The second principle is to ensure hukou reform takes effect in big cities, and not only medium and small-sized cities (Lu & Wan, 2014).The third principle pertains the need to lower entry requirements for gaining city hukou in big cities. The fourth principle is taking objective parallel reforms in public services and social security, an endeavour that will benefit the entire population in China, including rural dwellers. For instance, the central government could reform services such as education as local governments may find it challenging to invest in education for the rural residents, especially when such local authorities should increase educational expenditure (Lu & Wan, 2014).

In addition to the need to embrace the principles for reforming the hukou system in urban areas, urbanisation of mountainous regions in China should adhere to the sustainable development guidelines and policies. The concept of new urbanization is emphasised and expounded by Wang and Wang (2015), in their article New urbanisation: A new vision of China's urban-rural development and planning. According to Wang and Wang (2015), new urbanisation represents new government policy and guidelines of urbanisation, and it portrays a new stage in the development of towns and cities in China. The Communist Party of China's (CPC) National New Type Urbanization Plan 2014-2020 in the country has been embraced as a strategy for national rural-urban development, considering the average level of urbanisation across China has exceeded the set 50% in 2011 (Wang & Wang, 2015). The national urban development strategy has the potential to urbanise China, and this includes the underdeveloped regions in the country.

The policies and guidelines for urbanisation are crucial in helping the municipal administrations to regulate urbanization development. In particular, the policies influence and define different aspects that characterise urban planning. One of these is the way industry and agriculture relates where the government promotes the co-existence between the two. The guideline is in line with the desakota idea by McGee (2008) regarding the need for co-existence and intermingling between urban and industrial forms of land, settlement and agricultural land (McGee, 2008). The policies and guidelines also determine the ties between rural and urban and also among small, medium, and large cities (Wang & Wang, 2015).Wang and Wang (2015) further state that a combination among the policies, guiding regional ties, central local relations, and the background of development strategies in the country for the various regions acts as a driver for the urbanisation process. The article by Wang and Wang (2015) critique the "industry and agriculture combination policy" during the period of planned economy, noting that it only implemented a strategy that was dominated by industrialisation and production. In other words, there was a separation of industrialisation and urbanisation. From the 1960s to the Cultural Revolution, practices such as "sending intellectuals and urban dwellers into rural and mountainous regions" and "counter-urbanisation" caused a significant stagnation of the economy and urbanisation process.

Emphasis on coordinated and sustainable development by the Chinese government remains pertinent to the development of small cities and promotes local urbanisation of villages. Sustainable Development objectives are equally relevant to mountainous settlements in the country as it could support the idea of an alternative urbanity. The idea of a new urbanization strategy should also be implemented in developing mountain territories in China. The approach is characterised by emphasizing the sustainable and equitable features of contemporary urban design. Such include human-oriented, smart, green low-carbon, ecologically civilised and environment-friendly features (Wang & Wang, 2015). The strategy is in line with the idea of creating sustainable and liveable city regions by McGee (2008). According to Wang and Wang (2015), the designed features in urban regions are necessary for the coordination of production, synchronised modernisation of agriculture, living and eco-space, informationalisation, modern industrialisation and new urbanisation.

The National New Type Urbanization Plan (NTUP), the government's plan on urbanisation focuses on a path that is environmentally friendly and human centred. Taylor (2015) describes the urbanisation plan in concrete terms as involving urban hukou expansion, reforming rural land use and also infrastructure investment and development. The article by Taylor (2015) outlines the concept of "The China Dream" which emphasises a "moderately prosperous society."The people-centred urbanisation in China addresses the division between urban and rural regions and also within cities. It is also aimed at giving urban residency in a civilised and orderly way to peasants who have migrated to urban regions (Taylor, 2015). In the article, the author assesses the role of the China Dream in expanding urbanisation as well as promoting the efforts by Chinese government in maintaining social stability and boosting domestic consumption.

The Chinese Dream is founded on the projection that the future of the country is an urban one. It is estimated that within the next 15 years, about one billion Chinese, which makes up more than 70 per cent of the population, will be urban residents. It is for this reason that in March 2014, the Central Committee of the CPC and the State Council released the New Type Urbanisation Plan, which was the country's first official plan on urbanisation (Taylor, 2015). The China Dream can be interpreted in three intertwined ways. These are within the history of the country, as a driver to enhance domestic legitimacy and also as an approach to elaborate the concept internationally. Moreover, the China Dream is important in developing a comfortable lifestyle for the middle class and a moderately prosperous society. The concept of China Dream receives positive acknowledgment domestically, but its precise meaning is open to interpretation outside Chine, with some critics terming it as the "American Dream with Chinese characteristics" (Taylor, 2015).

The definition of China Dream includes the desire for major power status, rejuvenation of the country, and a military strong nation. In addition, it describes the wish for a prosperous future, a spouse, and a quality home; hence found in the urban region. Therefore, the National New-Type Urbanisation Plan completes the China Dream as its key aspects, i.e., urbanization and townisation, are expected to improve consumption and the domestic growth of the country now and in the future.

China's biopolitical project: 70% Urbanized Population Target:

The Making of a New Middle Class and transition from development & infrastructure investment-based economy to a Consumption-based economy.

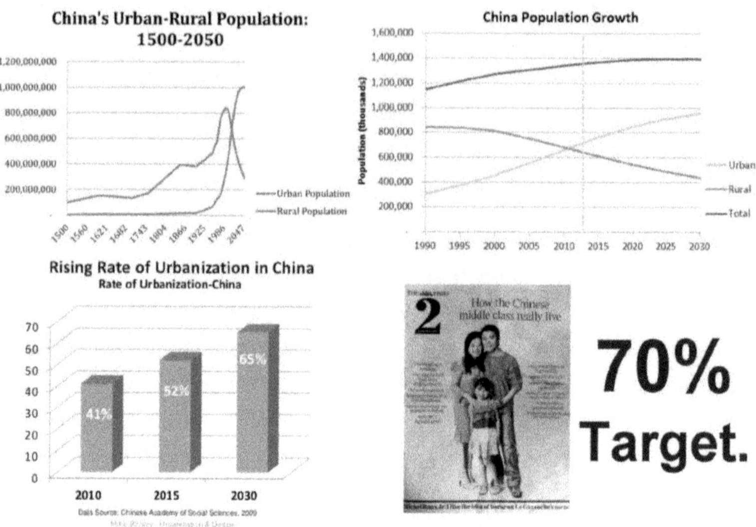

Fig. 1.12: The phenomenon of a middleclass in China is a relatively new concept. It is a society in the making, defining how to find livelihood strategies and how to coexist in established and emerging cities. Source: Various sources, The Times, UK.

According to Taylor (2015), a major priority of urbanisation is turning migrant employees into urban citizens, which requires fundamental change of the hukou system. Policymakers and scholars have increasingly blamed the system for the lower rate of urbanisation in China. Taylor (2015) asserts that future urbanisation in China will be characterised by growth in small cities and rural towns and not merely by the expansion of megacities. The author proposes that this should be accomplished when the Communist Party of China takes the cities to rural populace instead of moving the rural residents to the cities. Similarly, the extension of this idea into mountainous regions in China may develop further the notion of mountain settlements as a city.

Understanding the process of transition in China requires more than consideration of both political and economic institutions. For a full understanding of this process, it is essential to acknowledge that local cultural, political, and social resources also play a key role. The article From Mao to the Market: Community Capitalism in Rural China by Hou (2011) compares and contrasts three industrial villages in the country, showing different mixes along the organisational continuum necessary to attain a collective economy from market economy. In particular, Hou (2011) provides an alternative development model which integrates market production and distribution with redistribution and the creation of public goods based on group boundaries. The model proves applicable in the process of developing rural regions in China, and ultimately attaining a new type of urbanity in such regions.

As Hou (2011) explains, collectivism acts as the driving force towards urbanisation in China. While it is not the most effective form of production, collectivism addresses the problem where the state and market fail. According to Hou (2011), collectivism and rural capitalism prepare peasants in rural areas for the shift from an agrarian society to a society that is characterized by industrial endeavours. Moreover, collectivism brings resources together, making villagers better suited to compete in the market that often disadvantages the peasants. The fundamental justification for collectivism is that it avoids releasing peasants to try to make a living in urban areas, which would otherwise force them to reside in urban slums. In particular, collectivism keeps the rural peasants within the community and ensures they are employed; helping them transform the village (Hou, 2011). In such a collective setting, there is equal sharing of profits and risks among members of society; implying that such villagers become increasingly committed to improving the means of production for better outcomes.

Collectivism grants villagers the incentive of gaining acknowledgment and respect from other individuals, and this is similar to an "economy of regard." In this way, it is clear that the collective model remains valid in that it protects the underprivileged villagers (Hou, 2011). The protection is due to the potential to pool social, economic, and political resources. This provision protects a majority of peasants who are not prepared or ready to engage in market competition. The hukou system continues to constrain peasants as even those who accept to leave their land and become workers are highly limited by social mobility.

Similarly, such rural residents who move to urban areas can only take on stressful and lowest-paid jobs. According to Hou (2011), the collective model grants peasants a safety net; hence, they do not depend on individual abilities. In this way, this alternative model accelerates rural industrialisation and urbanisation and at the same time, ensures that development of health care, education, infrastructure, and other essential public goods is prioritised. Remote rural mountain settlements in China would, therefore, benefit tremendously from the collectivism and community capitalism, as this will accelerate the rate of urbanisation. The paper Geography, economic policy, and regional development in China by Démurger, Sachs, Woo, Bao, Chang, and Mellinger (2002) assessing the potentials of urbanising rural settlements in China could also benefit greatly from understanding the impacts of policy and geography on growth rates of different regions. As the authors elaborate, Chinese civilization started in 2000 BC in the northwest, which is approximately 1,000 kilometres from the coast (Démurger, Sachs, Woo, Bao, Chang, & Mellinger, 2002). The underlying reason as to why the northwest was China's economic centre is that in ancient times, land-based trade and increased agricultural endeavours, which happened in the northwest, were crucial compared to sea-based trade. In addition, most of the country's international trade happened through the Silk Route, passing through China's north-western corner. On the contrary, the south-eastern coastal region, which is the location for Guangdong and Fujian, was uncultivated and had a sparse population (Démurger et al., 2002). While there were favourable conditions for agriculture in the southeast, farming was underdeveloped due to subtropical diseases, i.e., malaria as well as the impacts of high temperatures which caused reduced labour productivity (Démurger et al., 2002).

The article by Démurger et al. (2002) expound on the growth of the coastal region of China. According to Démurger et al. (2002), the so-called preferential policies that coastal provinces receive are mostly deregulation policies that enable them to integrate into the international economy. In this light, it becomes superior to decrease regional disparity through the extension of the deregulation policies to the interior provinces compared to the results of re-regulating the coastal regions (Démurger et al., 2002). Reducing disparity in regional incomes proves a reality in geographically large countries, and this problem has multiple and complex causes, and it is for this reason that the article by Démurger et al. (2002) seeks to identify the role that geography and policy play in bridging the regional backwardness of land-locked territories.

When addressing the problem of regional disparities, it is pertinent to understand that development strategies that have been successful in some nations are not guaranteed to produce similar salubrious outcomes when applied in other national contexts. In this sense, it is essential to point out that not all development strategies would be successful in rural regions of China. Démurger et al. (2002) offer the example of when China opened certain pockets for foreign direct investment (FDI), the emerging Special Economic Zones (SEZs) rapidly produced vibrant platforms for export but on the downside, led to backward linkages with the surrounding hinterland. The lesson from this mechanism is that "the time-tested, effective growth policy package" for an economy in the coastal region, as well as its modifications to some extent, barely works for an economy in the interior areas (Démurger et al., 2002). Therefore, it is evident that the development policies that prove suitable for coastal regions are not guaranteed to work for rural settings.

Geography proves a particular factor when assessing the success of development policies. According to Démurger et al. (2002), two geographical divides characterise the wealth of nations. The first divide focuses on differences in ecological characteristics, i.e., the tropical versus temperate zone while the second geographical divide describes the variations in the potential of a region to engage in international trade, and under this consideration, there is the differentiation between coastal and interior areas. Both divides apply when evaluating the possibility of 'rurban' mountain settlements in China. For instance, after a long period of constant expansion towards the south, the Chinese empire halted where the tropical zone begins (Démurger et al., 2002).

Geography influences regional income levels through two major channels, namely agriculture and international trade and FDI. As Démurger et al. (2002) inform, physical location is a crucial determinant for development as it is associated with such factors as cost of water transportation that makes it easy for coastal provinces and regions along navigable rivers to produce manufactured exports. The author also notes that a steep landscape is associated with increased agricultural costs (especially cultivating crops) and transportation of goods. The possible solution in such mountainous regions is infrastructure construction, i.e., tunneling and terracing, but this also proves expensive. Therefore, developing mountainous regions in China is particularly challenging due to two major factors, i.e., geography and limited land for farming.

Another particular endeavour that proves significant to 'urbanising' mountainous regions in China pertain income diversification. Economic reforms in rural China have informed the diversification. In a second article by Démurger with Fournier and Yang (2010) Rural households' decisions towards income diversification: Evidence from a township in northern China the authors treat the idea of income diversification, noting the role of economic reforms in rural China. In particular, the authors analyse the reasons driving rural residents and households to engage in diversifying their sources of income. According to Démurger, Fournier, and Yang (2010), economic reforms in rural regions of China grant rural residents the opportunities to diversify off-farm and within-farm endeavours. Engaging in activities increases the income of rural households. The launch of economic reforms in China from the end of the 1970s led to significant transformations in productive endeavours in rural regions.

Diversification of China's rural economy was key to a steady rate of urbanisation of these regions. According to Démurger, Fournier, and Yang (2010), agricultural endeavours overwhelmingly dominated the rural economy in China during the pre-reform period. Grain crops were the major produce as they accounted for over 80 per cent of the entire cultivated area. Economic reforms availed opportunities and incentives for rural households, enabling them to diversify activities within farm and off-farm. Initially, there was the Housing Responsibility System that caused the breakdown of the People's Communes (Démurger, Fournier, & Yang, 2010).

Similarly, this enabled rural homes to make decisions about their economic endeavours. Changes in prices and growth of free markets during the same period also offered incentives to engage in profit-oriented endeavours. In December 2001, China joined the World Trade Organisation (WTO), and this has increasingly accelerated structural changes as currently, labour-intensive endeavours characterise agriculture, compared to the traditional land-intensive production of grain (Démurger, Fournier, & Yang, 2010). The modern-day diversification entails the production of such crops as vegetable crops, fruits, animal husbandry and also industrialisation of rural regions. The transformation is promising in endeavours to urbanise underdeveloped mountainous areas. In particular, rural residents should embrace income diversification, which will result in considerable growth of the rural economy. For example, production of different crops will favour industrialisation of such rural regions, and ultimately, peasants could increasingly enjoy access to benefits that urban residents receive, by an increase in income.

In practice, engaging in income diversification requires persistence and determination. According to Démurger, Fournier, and Yang (2010),land scarcity and absence of any incentive and comparative advantage to engage in agricultural endeavours characterise many areas of western China, mainly villages in mountainous regions. The major off-farm activities in these regions are migration and tourist development. Conditions for success in developing the rural regions in China depend on the potential to increase access to off-farm endeavours for all rural residents, especially for households with little land, human, and monetary access. In this light, it is essential to develop off-farm efforts such as promote local rural industry, migration and tourist industry as this will assist in reducing rural poverty and increasing rural income (Démurger, Fournier, & Yang, 2010). The long-lasting impact would be sustainable development of such regions.

In their article, Sensitivity of livelihood strategy to livelihood capital: an empirical investigation using nationally representative survey data from rural China, Xu, Deng, Guo, and Liu (2018)expound on the idea of diversification of income sources in Chinese rural regions. Xu, Deng, Guo, and Liu (2018) divide the surveyed villages into three categories namely hilly, plain and mountainous villages. The authors acknowledge the increasing gap in income between the agricultural and industrial sector, further noting that many nations continue to experience significant transformations in the structure and income of rural households. In particular, Xu, Deng, Guo, and Liu (2018) note that the increased rate of urbanisation and economic growth in the country leads to rise in the ratios of part-time labourers and off-farm labourers compared to the total household labourers from 12.07% and 19.95% in the year 2008 to 19.05% and 28.71% in the year 2016, respectively. Similarly, the ratio of the population who practise farming compared to the figures for household labourers has reduced to 38.35% in 2016 from 51.16% in 2008 (Xu, Deng, Guo, & Liu, 2018). The changes are further evident as the ratio of off-farm income to family income has risen to 61.65% in the year 2016 from 48.84% in the year 2008. In this sense, it is apparent that diversification of income activities has been on the rise, with a majority of rural residents increasingly engaging in activities other than agricultural production.

Migration of workers and the shift to family income structures has been instrumental to the differentiation of rural homes. These changes play a crucial role in attaining economic growth in rural areas, which ultimately leads to gradual 'rurbanisation' in such regions. Nonetheless, it is essential to point out that some rural residents still engage in agricultural production as their fundamental means of livelihood whereas others consider off-farm activities as the primary sources of livelihood (Xu, Deng, Guo, & Liu, 2018).The authors point out that research on the livelihoods of rural households remains a critical and significant topic in various disciplines such as geography, economics, ecology, management, and others. Different rural homes in various villages have similar structures for livelihood asset and selections of earning strategy. The findings of the study by Xu, Deng, Guo, and Liu (2018)indicate that most rural households greatly value human capital as their livelihood asset. In line with diversification of income, rural households select off-farm work as the livelihood strategy.

Few individuals engage in on-farm work, and for this reason, there is emergence of activities often found in urban regions, such as industrialisation (Démurger, Fournier, & Yang, 2010). Residents in mountainous territories prefer financial and human capitals, and for this reason, efforts to urbanise such regions would greatly benefit from policy support from the government. Besides, the state-party should motivate excellent and experienced migrant labourers to return to their rural hometowns and engage in entrepreneurship (Démurger, Fournier, & Yang, 2010). This endeavour will encourage other rural farmers to adjust their livelihood strategies, which will lead to considerable economic growth; hence boost sustainable urbanisation at grassroots level.

Development of e-commerce in China proves promising towards urbanisation of rural regions in the country. E-commerce in China has developed at a considerably high rate in the past few years. The article by Li (2017) expands on this development, noting that with the expansion of e-commerce since the 2000s, "Taobao Villages" have appeared in rural regions of China. According to Li (2017), the "Taobao Villages" are clusters of e-commerce endeavours since the mid-2000s, and these are after the Taobao or e-commerce platform that Alibaba owned. With the spread of Taobao Villages across China, there are increased expectations among the Chinese as well as the international community regarding the role e-commerce will play in developing stagnant rural economy, ultimately alleviating poverty in rural areas of China and also outside the country (see Fig. 1.13). Li (2017) explores the role of Taobao Villages in transforming rural society, the economic sustainability of such villages, the role of party-state in the development of the Taobao Villages as well as how the villages have changed the dynamics of rural governance. Overall, the article by Li (2017) informs that Taobao Villages can significantly benefit peasants and the rural population in China. 'Private' Internet enterprises largely occupy the internet sector in China. The state is least involved in this sector, compared to its involvement and domination in other sectors of strategic value, for instance, electricity, telecommunications, petroleum, and finance. Nonetheless, it remains significant to point out that the 'private' sector maintains important ties with the state in the provision of internet. Various internet giants have near-monopolistic advantages in their respective markets, for instance, Alibaba dominates e-commerce, Tencent in social media, whereas Baidu is the most influential online search engine (Li, 2017). The dominance of these internet giants transcends the country's border. In this light, China is increasingly becoming the biggest e-commerce market in the world, implying that the advantages of such e-commerce will transform every part of the country.

Fig. 1.13: New City : Taobao Village. Shenzhen Bi-City Biennale of Urbanism\Architecture (UABB) 2018. Video by Liam Young and Alexey Marfin. Soundscape by Forest Swords. Commissioned by the Shenzhen Biennale.

The government, in conjunction with private Internet enterprises, continue to encourage the development of e-commerce in rural regions. In particular, e-commerce is integral to attaining a prosperous society by the year 2020 and beyond via "new-type urbanisation" and incrementally alleviating poverty in rural areas (Li, 2017). However, the efforts by the government face considerable challenges such as the digital divide between rural and urban context, which is evident as rural regions have lower Internet penetration and compromised levels of applying digital technology. The government could establish policies such as Village Informationalisation Programme (VIP) and such others to bridge the existing rural-urban rift. In the past decade, the development of rural e-commerce has produced rural entrepreneurs who earn a livelihood through online trading. This is a form of income diversification, which also plays an important role in urbanising rural areas (Xu, Deng, Guo, & Liu, 2018).The entrepreneurs are also involved in the rise of Taobao Villages and also in creating huge markets for rural e-commerce. With the policy support of the government and considerable investment from Internet and e-commerce giants such as Alibaba, Taobao Villages have increased rapidly creating the new rural e-commerce service system which is pivotal in urbanising the rural areas (Li, 2017).

China has the highest number of Internet users of any nation and the biggest e-commerce market in the world. These numbers are estimated to be more than twice the America's size. Despite the large population of Chinese using the Internet, many more, i.e., hundreds of millions are yet to embrace online presence (Fan, 2018). The prediction by analysts is that the online retail market in China will be double in the next two years. The promising part is that the growth is estimated to come from smaller cities, i.e., third and fourth-tier cities. Similarly, the vast rural hinterland of the country will greatly contribute to the increase in the number of Internet users (Fan, 2018).The Chinese government is increasingly involved in developing infrastructure programs to raise the standards of the rural regions. At the same time, the private sector plays a significant role in developing the areas. The article by Fan (2018) explores the role of giant e-commerce companies such as JD.com or Jingdong (as it is commonly known in Chinese) in increasing economic growth in rural areas of China (see Fig. 1.15).

Venturing into uninhabited regions including mountain regions, could increase the potential for economic growth, further increases the success of giant companies. For example, a major growth strategy for JD is to market its services to over two hundred villages around the Wuling Mountains. JD.com has a policy of employing local representatives who rely on their social ties for the benefit of the organisation. Rural residents are regions are characterised by practices such as farming, foraging, barter trade, and also exchange of favours (Fan, 2018). Therefore, involvement in any business endeavour in such rural areas requires prior knowledge of the activities of rural residents. Delivery personnel working for JD are conversant with the existing social networks, and this knowledge acts as a significant asset for firms operating in such regions. In this light, urbanisation of remote rural mountain settlements in China is possible through operations of giant e-commerce firms in such regions. The personnel employed are assured of a better source of livelihood, in comparison to other activities such as farming. Besides, the business knowledge gained is suitable to transform such rural workers into entrepreneurs.

The extension of delivery networks enables remote communities to independently offer local agricultural produce to larger cities.

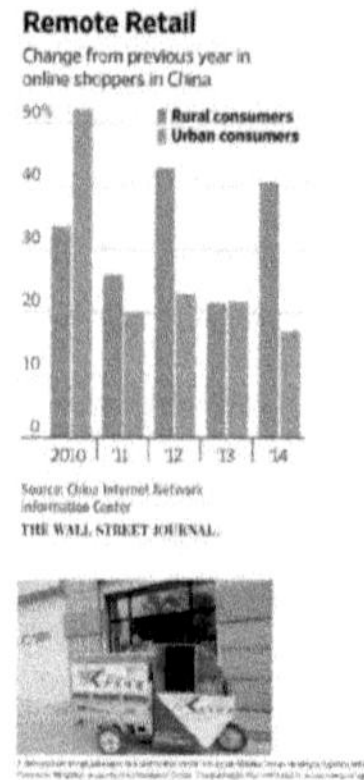

Image Source:Zuo Jing in "Bishan Magazine," 2014.

Fig. 1.14: Online retailers as Alibaba expand their e-commerce network into the rural areas of China for further growth. Sources: The Wall Street Journal; Gillian Wong and Loretta Chao, August, 2015.

Novel networks of mobility and connectivity innovation.

The proliferation of new types of networks and connectivity is rapidly extending into mountain areas both promoted by Top down governmental support, Chinese corporations, agricultural collectives and Bottom up local initiatives by private families.

Fig. 1.15: Large Chinese corporations are setting up drone delivery networks in rural areas. While hiring local residents and hiring them for drone delivery services, this expansion into remote villages is promoted by top down government support. It is relying on knowledge of the existing social networks and linking them to large national scale distribution systems. Eventually, the meshing together of existing of social communities with new networks can help to disseminate urban knowhow, better information sharing, and easier access to knowledge, education, medical care and social welfare support. Source: Illustration by Jun Cen, Jiayang Fan, The New Yorker, July, 2018.

1.4. Hypothesis and Research Questions:

Hypothesis:

Alternative Urbanities in the mountains will resist resource exploitation by mega-city clusters. Developing self-reliant micro-ecologies will be crucial for attaining independence.

Bottom Up diversification of livelihood activities will better withstand a global economic downturn. Welfare and Environmental Infrastructure in mountain areas will support diversification.

By embedding state-of-the-art knowledge in transects more equitable livelihoods and revitalized ecologies can transform mountain territories.

Research Question One:

The objectives of China's National New Type Urbanization Plan 2014- 2020 include a balanced approach to rural – urban development. In what way can the livelihood strategies of individuals be adapted for a long-term sustainable transformation of mountain settlements beyond 2050?

Research Question Two:

Will Alibaba's rural expansion programs lead to emancipated settlements in impoverished mountain regions in China? Alternatively, what Repertoire of Interventions is needed for an independent Alternative Urbanity in the mountains?

Research Question Three:

In China the rural-urban transect method does not have a long tradition, compared to the European and North American planning practices. In what way can the transect approach be applied in fragmented rural-urban territories anticipating a people-oriented, new type urbanization phase?

Part Two

Contextualization.

The Maoist Cultural Revolution in China 1966-1976, the beginning of a gradual overlapping of the countryside and the city, and the development of rural-urban settlements.

Fig. 2.01: The poster gives greater visibility to farmers, factory workers and the army. Urbanites are encouraged to embrace a rural work ethic and livelihood strategies. Poster: 'Chairman Mao Zedong Vast World of Accomplishments'. People's Fine Arts Publishing House, 1967. Source: Etsy Vintage Reprints.

Part Two: Contextualization.

2.1. Introduction: Overview of urbanization phases in China.

The following section provides a concise overview of the main urbanization phases in China since the late 1940s (see Fig. 2.02). It outlines the key objectives and agendas for each of the phases including the current New Type of Urbanization Plan. Furthermore, this section presents some of the current urbanization tendencies in China and indicates possible directions for future urbanization phases. Although there are multiple variations of the trajectories to the process of urbanization in China, broadly-speaking four groupings can be described. An early urbanization phase is a pre-economic and pre-industrial development of settlements in rural and coastal regions. The second type of urbanization is driven-by top down political and military power, the development of infrastructure and new settlements mainly serves national defence and security objectives. A third type of urbanization is evolving out of the emergence of industrial and mining towns. The forth type of urbanization path is a combination of industrial cities attracting foreign direct investment, local municipality participation, and the transfer of new knowhow and technologies into rural areas.

Phase One: 1949-1957 Industrialization-led Urbanization.

The first phase of urbanization from 1949 emphasised the development of predominantly heavy industries. Approximately, one hundred-fifty-six industrialization projects were supported by the USSR during this phase. The migration of the population in China was controlled by the government. Citizens were encouraged to move to new employment opportunities in factories and on construction sites. It is an urbanization process where the development of industries was supplemented with the provision of housing projects and social welfare programs. This phase of urbanization saw a relatively high growth in urbanization and industrialization rate. During the 8 year period at least eight industrial cities were formed with an urbanization rate of 10.6% to 15.4%, or a growth rate of 0.53% annually.

Phase Two: 1958-1977 Great Leap Forward and The Cultural Revolution.

The Great Leap Forward Campaign from 1958 to 1960 was a Soviet-style industrialization period, where the development of industries was predominantly funded by proceeds from the agricultural sector (see Fig. 2.03). This phase saw an urbanization rate of 15.4% to 19.8%. Unfortunately, the Great Leap Forward became disastrous to the agricultural communities, pushing farming families to subsistence level or hunger. Consequently, this development phase resulted in the Great Chinese Famine from 1961 to 1963. During this period, the urbanization rate was between 16.8% to 19.8%.Temporarily, the country was also experiencing the deterioration of Sino-Soviet relationships, contraction of the construction industry, loss of employment opportunities, and a reduction of the urban population.

During this crisis period no new major city was developed. Because of the difficult Sino-Soviet relationships during this period, the Chinese government launched the development of Third Front Industries. It emphasized industrialization without urbanization. To protect industries from foreign military powers, factories were moved into the mountains and caves. It resulted into a de-concentration of industries away from the coastal regions. The Cultural Revolution from 1966-1976 promoted the migration from cities into mountain territories. It had an urbanization rate of approximately 17% between 1962 and 1977. Part of the Cultural Revolution program was to 're-educate' younger urbanites and intellectuals by sending people to rural areas and the mountains (see Fig. 2.04).

Four main phases of Urbanization in China since 1949 - 2010.

Forty years of an export-based national rejuvenation program.
The unintended consequences include Uneven Urbanization, Distortions in the Urban system and Inequality.

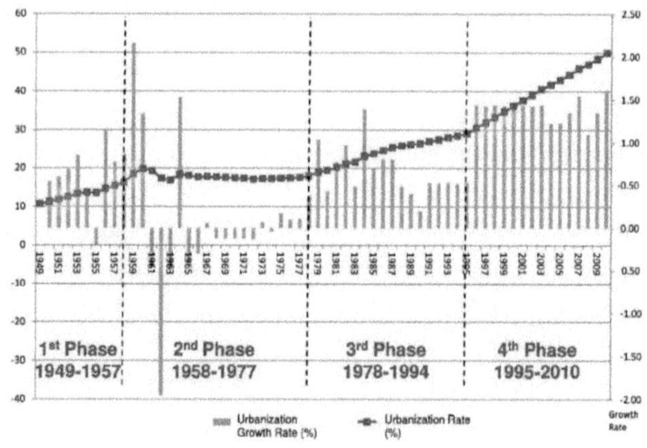

Fig. 2.02: The diagram shows the Urbanization Rate & Urbanization Growth rate in four phases of Chinese history. Source: Kuangdi, X., 2106, Scipedia, SL, 2018.

Great Leap Forward (1956-1960)

Fig. 2.03: 'The people's communes are good ' by Rui Guangting,1958. The artwork for schools illustrates the ideal people's commune, with a healthy harvest, communal facilities for sustenance and sanitation, a centre for the older population and the people's army. Shanghai educational publishing house. Part of the IISH, Stefan R. Landsberger Collections & Anchee Min, Sinological Institute Leiden University.

Phase Three: 1978 – 1994 South-East Coast Urbanization Priority.

The Household Contract responsibility system was introduced to solve the food production crisis temporarily. Additionally, this phase started to prepare for urbanization during an incremental liberalization of the Chinese economy. To rejuvenate the country and to plan the transition from an emerging economy into a global superpower, the Chinese government pursued an internationalization of the economy.

In 1979 Special Economic Zones such as Shenzhen, Zhuhai, Xiamen, Shantou, in the two provinces Guangzhou and Fujian were established as platforms for international trade. By, 1984 additional fourteen special economic zones were created along the south-eastern coastal zone. Additionally, Coastal Special Economic Zones in the Pearl River Delta, the Yangtze River Delta, and the Fujian Delta were set up in 1985. The Shanghai - Pudong area followed the same economic model since 1990. Furthermore, in 1991 twenty-six National High-Tech Development zones with four ports opened for trade in the Northern coastal zone.

Within one year in 1992, Deng Xiaoping initiated an economic development program creating new platforms for trade in coastal cities. Starting in the Yangtze River and extending trade liberalization into the inland this development program involved eleven capital cities in the inland of China with international business networks.

Up to the Mountains and Down to the Countryside Movement was a policy during the Cultural Revolution in China.

Fig. 2.04: "Educated youth must go to the countryside to receive reeducation from the Poor and Lower Middle peasants!," 1969. Source: part of the IISH / Stefan R. Landsberger Collections, and Ancee Min.

The overemphasis of the development of Coastal Cities in China has led to unbalanced growth. Increasingly, landlocked and Mountainous territories with limited access to the sea being discovered for the next phases of urbanization in China.

Fig. 2.05: A map of Greater China with the Coastal cities emphasized as the main population centers driving a sea-based, export-led economy since the late1970's. A turn to the interior provinces will drive economic growth and urbanization through the development of medium and smaller sized cities. Image sources: by the author, CEIC, WTO.

China Develops Capacities as a Global Super Power (1989-present)

Fig. 2.06: The Chinese poster presents advancements and innovation in science, technology, space exploration and genetic research. Credits: Author unknown. Publisher: Shanghai renmin meishu chubanshe, Feb. 2001. Part of the IISH / Stefan R. Landsberger Collections, Sinological Institute Leiden University.

Phase Four: 1995 – 2014 Market Economy and Urbanization.

The Oskar Lange's economic theory of Market-Socialism partly influenced China's economic policy and modernization of the country. Additionally, the need to acquire international standard technologies, scientific knowledge and management expertise shaped this phase of urbanization to participate in the global economy. Gradually, since 1988 land use reforms enabled the supply of agricultural land for urban development. It gave farmers more freedom to develop mixed uses of agricultural-industrial land together with the development of housing and tourism related dedication of land (see Fig. 2.06).

In 1994, the government introduced housing reforms allowing for a market-driven development of residential projects. From 1998, the government promoted the monetization of land, or leasing of land by rural municipalities for urban real estate development. Since 1994 the government also emphasised motorization of rural citizens by promoting the acquisition of motorbikes and automobiles. The objective was to increase mobility of the population and to enhance accessibility of settlement outside of the main urban centres. Furthermore, tax reforms were introduced to stimulate economic growth in support of new enterprise formations. The formal integration of China into the Global Economy was formalized by the joining of the World Trade Organization in 2001 (see Fig. 2.07).

New Type of Urbanization Plan 2014-2020.

The National New Type of Urbanization Plan is the first official urbanization plan in China. Its objective is to advance a people-friendly policy to urbanization taking into consideration global sustainability challenges including climate change, food security, environmental pollution, and the preservation of ethnic identities. Additionally, a balanced approach to urban – rural development reducing poverty in rural areas is advocated. Greater coordination, to avoid an overcapacity building in some sectors and a duplication of industrial production between provinces and competing cities within the country requires further optimization. Due to the relatively short time span for the implementation this plan, further adaptations of the aspirations of this urbanization plan can take into consideration the mountain territories as part of an urbanization program.

New Type of Urbanization Plan in China 2014-2020:

A balanced people-oriented urban-rural development, food independence, poverty alleviation, Environmental Sustainability and preservation of ethnic identities.

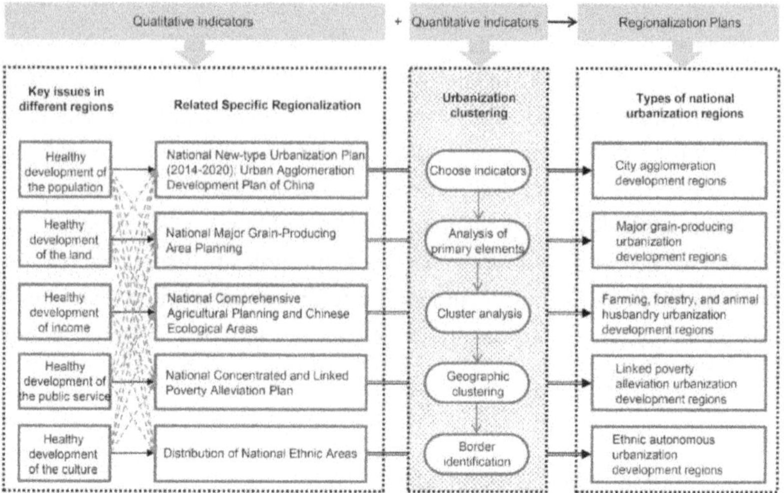

Fig. 2.07: National New type Urbanization Plan (20142020): For the first time in 2014, an official national urbanization plan was issued by the Central Government. The policy, initially, for five years seeks to address coordinated regional development. It aims to tackle unintended consequences to human and social issues as a result of urbanization. Furthermore, its objectives are (1) Balanced Urban-Rural development, (2) Coordinated Regional Development, (3) Environmentally friendly and Sustainable Urbanization, (4) Biodiversity Preservation, (5) Food Independence, (6) Poverty Eradication, and (7) Preservation of Ethnic Identities. Credits: Fang, Ma et Wang 2015.

Future Aspirations: Retrofitting of Cities and Mountain-Urbanity.

First and foremost an improvement to the living conditions in polluted cities is put forward in the "Blue Sky" policy action plan for 2018-2020. It will likely be followed by subsequent air-pollution action plans as cities continue transform and a clean energy transition plan will require more time for implementation. Furthermore, the settlement of migrant citizen as permanents citizen through reforms of the household registration system could gradually give people more access to public welfare system, education, health care and affordable housing. For an urban development phase beyond 2050, the accelerated construction of urban developments since the 1980's which resulted in several derelict, underutilized buildings and public infrastructure will be best addressed by an adaptation to retrofitting existing buildings, change of use and conversions of the as-built environment. Part of a future urbanization approach looking towards 2050, may be the planning for alternative cities in rural and mountain territories with the same access to information and know-how as in urban areas. A shift away from the dominant mega-city network emphasis to urbanization may alleviate some of the social and environmental problems found in mega-cities cities. By re-settling people in medium and smaller-sized cities, a de-concentration of major population centres may be achieved. Smaller population centres with diversified local economies and independent generation of food and energy resources could improve the uneven development of territories and support a greater autonomy of smaller-scale settlements.

Classification of city size by Chinese standards:

Mega-Cities, Big, Midsized, Small and Big Town.

The number of Cities with populations of +/- One Million inhabitants will increase by 2030, and beyond.

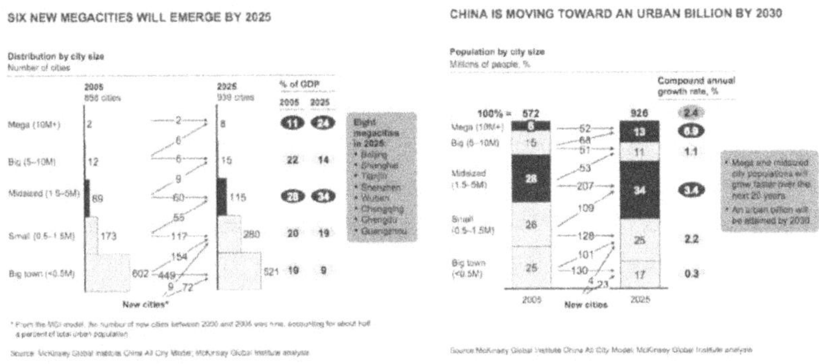

Fig. 2.08: In the next phase of urbanization in China, growth will be mainly in the number of Medium size cities (1.55 million people), Small size cities (0.5 to 1.5 million people), and Big Towns (> 0.5 million people). China counted approx. 858 cities in 2005. By 2025, the number of cities could rise to 939, out of which 8 will be Megacities with over 10 million urban citizens each. Source: McKinsey & Co. report "Preparing for China's urban billion" March 2009.

2.2. Mountain territories & Metropolitan agglomeration:

Rethinking Urbanity:

The unintended negative consequences of accelerated urban growth provide an impetus to reactivate mountain territories as habitats for a different way of living. Mountains as new type of cities may help to lessen the burden imposed on the environment and communities in developed population centres. Innovations in the modes in transportation have the potential to give access to spaces for settlements, which were previously largely uninhabited or uncultivated land. Experimentations with more efficient smart farming and food production technologies could be brought to remote areas to modernize smaller-scale family farms in mountain areas. This could free up labour from traditional farming practices and thus reduce the burden on elderly and mothers often left behind in mountain farm households.

Modernized farming operations in mountains could facilitate greater diversification into non-agricultural economic activities assisted by the access to the internet. Cleaner energy sources in mountain areas could help to reduce the cost of operating a small farm, and thus allow for better chances of reinvesting any potential disposable income. The further the progress of digitization in mountain settlements with an improved access to information, and exchanges of knowledge with people in urban areas can eventually enhance the education of local people, and strengthen the self-sufficiency of remote communities.

The democratization of digital fabrication technologies, distribution of products and services on on-line platforms could help to form new productive ecologies in the mountains villages. With shifts of the centres of economic activities in regions over time, the potentials of mountains as a settlement space is gaining more attention. Offering more sustainable and liveable environments, the advantages of integrating mountains as part of larger metropolitan regional strategies may alleviate some of the urgent social and slow-burning environmental stresses (Rodin, 2014) in densely populated urban areas.

While often overlooked, mountain territories as an ecological and spatial resource play a permanent role in regional rural- urban systems. Mountain areas carry vital environmental functional capacities such as preserving biodiversity, the provision of clean water, energy production, the absorption of pollutants, air purification, land for farming and dwelling (see Fig. 2.09).

As spaces for recreation and wild life protection, mountain territories with scenic landscapes are destinations for tourism, sports and the hospitality industry providing a source of revenue for local residents living in mountains. As a security provision, in the case of foreign military invasion, the strategic concealing of military training centres, airports hidden in mountain caves and defence related research laboratories are secret and yet common uses of mountain regions in several countries. Given such centuries-old functions of mountain territories, an urgent rethinking of the possibilities for settlements in steeper terrain is increasingly being considered. In particular, during a period of urbanization, when development is consuming agricultural land, and recent ITC technologies and communication connectivity entering remote regions offer new options for residing outside of large cities.

Challenges for the transformation of mountain settlements:

Some of the inherent difficulties for the development of mountains are its topographic features, the limited fertile or arable land, a sloped terrain above 10 degrees, and the issue of land-locked regions with limited access to the sea or navigable rivers. Other challenging factors to consider are the higher costs of transportation and possibly unfavourable climatic conditions for farming. For instance, an arid or tropical climate may lower the productivity of agricultural industries. High or too low temperatures, subtropical humidity levels, pathogens and infestations may also lower the productive energy levels of the agricultural labour force working outdoors. The causes of regional backwardness may also be attributed to the inflow of foreign capital predominantly into centres of economic activity and established cities, as well as the outflow of migrants seeking employment or education in urban areas. Advantages for settlements in mountain areas are the lower cost of living and to some extent subsistence farming for local families. Most importantly, healthier environmental conditions including cleaner water, air and soil are benefits increasingly diminishing in larger metropolitan cities.

Ecological, Regenerative, Habitat and Economic functions of Mountain Territories are recognized as integral to the systemic stability of urban systems Larger spatial configurations include metropolitan regions, the Intermediate Territories in-between and the mountains.

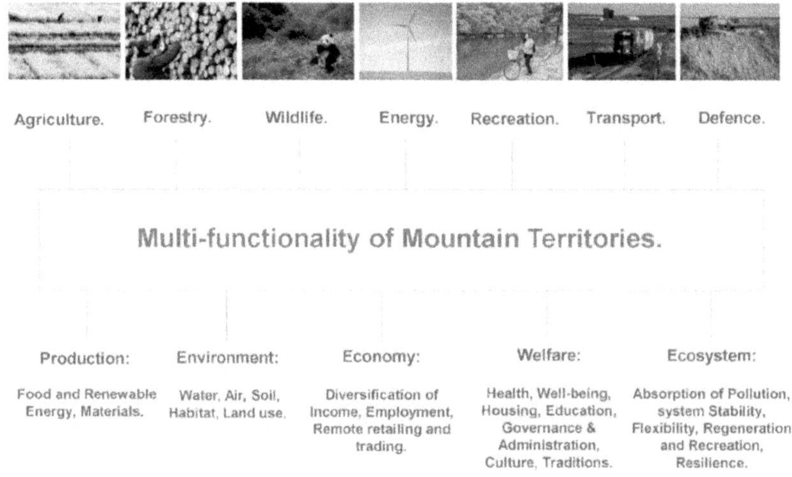

Fig. 2.09: The diagrams presents some of the known functions of mountains, as new forms of urbanity emerge new ecologies become embedded into the mountain regions. Source: Hans Vejre, M. Antrop, U. Mander.

The task of reducing the per capita income gaps between mountain territories and urban areas must be addressed by investment into education and better transport infrastructure development. Partly, income inequalities in the hinterland can be attributed to insufficient contacts with external sources of funding for industry development, upgrading of village roads and restoration of dwellings. To reduce poverty in the mountain areas speeding up economic growth in a sustainable manner is a difficult balance to maintain. It involves making difficult choices between potential short-term economic gains and the long-term objectives avoiding human and ecological damages. An ecology-friendly development strategy and infrastructure building, which puts value on coordinated and comprehensive planning takes more time to prepare compared to short-term development thinking not considered holistically. This would include a development program appropriate for the specific local geography and regional climate. A steeper topography raises the cost of cultivating land and the transport costs of agricultural produce. Geographical obstacles to some extent can be overcome by landscape design and belowground tunnel construction. However, the higher capital investments for the construction of terraces, excavation of earth and infrastructure development make mountains more difficult for human settlement. The setting up of new mixed agricultural-industrial operations has limitations to the potential of upscaling businesses due to the smaller size populations living on sloped terrain. The exodus of the younger people from mountain areas to cities leaves mothers with children and the elderly behind in the villages. This is limiting a supportive division of labour affecting the productivity and revenue generation. An over specialization of the rural labour force in mountains is better minimized to cultivate diversification. Unfortunately, a participation in international supply chains for products often forces businesses to specialize in a narrow set of manufacturing activities or farming produce. The expansion of local markets and participation in international trade is inhibited by the difficulty to attract foreign capital for physical infrastructure investment and development of industries.

Recommendations for settlements in mountain areas:

Despite the topographic features of mountain areas, the geographic distance from metropolitan centres can be an advantage, as limited accessibility can shelter from the competition with larger cities. The resettling of labour resources into low density dispersed settlement structures in steeper terrain can be made attractive as mountain areas can be more healthy habitats for living than overcrowded and polluted urban areas. The retaining of new communities in mountain valleys may be supported by innovating and modernizing communication technologies and informatization (see Fig. 2.10-2.12). Few specialized local product and services may be offered through global trade networks. For mountain areas to participate in export-oriented markets, rural enterprises must be equipped with digitized technologies and new mobility systems. Establishing exchanges between developed cities and self-sufficient mountain territories will initially require the building of resilient and financially more independent communities. The enhancement of local economic ecologies by the consolidation of smaller businesses into rural- industrial enterprises could form new synergies. The economic independence of mountain communities would imply, local livelihood strategies of people are adaptable to multiple states of equilibria. In other words, people are prepared to switch between different income generating activities and have diversified ways of supporting their families. The investment in environmental infrastructure, clean energy, ITC communication, and provision of low-cost transportation can be catalytic interventions for an economic shift, the improvement of agricultural productivity and community well-being. Similar positive impact investments in mountain areas may be initiated in existing mountain valley communities and then gradually be extended into steeper topography regions. The advantage of sheltering facilities in mountain caves should be used to set up and relocate sensitive intellectual property research incubators for new knowledge creation in a protected environment.

The idea of mountains as a custodian topography can be further considered as a competitive advantage for setting up of a new type of urbanity. To this end, human capital formation in a new type of knowledge community will be indispensable. For early stages of a transformation process, the transfer of knowledge and intellectual capital from metropolitan areas will be needed to enable the development of 'off-shore' research institutions. Gradually, mountain communities will adapt to behaviour, which may be equal to the behaviour urbanites in larger cities. People will have the same access to information, goods and services, while protecting the local environmental and knowledge capital. New type of innovation hubs and research centres in the mountains oriented towards environmental remediation and regeneration of the ecology could create future-proof industries. Ideally, this would be accomplished by securing renewable energy independence, protecting food security, community health and wellbeing. The embedding of new economic activities may include science, technology, education, health services, and sustainable tourism while maintaining the local ethnic identity.

'Agriculture 2.0': Precision farming and managing Environmental Variability and Market-price volatility of agricultural produce. Technology assisted optimization of Agricultural productivity, and detection of plant pathologies.

Fig. 2.10: As coined by IBM's Research Precision Agriculture, 'Cognitive' Precision farming technologies are designed to coordinate the timing for crop harvest informed by weather data and the commodity price volatility. Source: IBM Research Precision Agriculture, Kevin Lopez Alvarez, 2017.

Mountains topography utilized for installation solar panels for clean energy generation.

Fig. 2.11: Clean Energy farms in the Chinese countryside. On the interim, due to recent US trade tariffs, the installation of Renewable Energy infrastructure may be scaled-down. Source: Eric Ng, SCMP, March, 2018.

Smart farming and the application of new technologies in rural areas.

Fig. 2.12: The use of drones for spraying of pesticide in China's Shanxi Province. Source: Various Sources: Xinhua, Cao Yang; News.cn, Global Times, 2017.

2.3. Opportunities and Challenges of the Dongjiang River Basin & the surrounding Hakka Territory in Guangdong Province.

Introduction:

The following section discusses opportunities and challenges facing the Dongjiang River Basin & the surrounding Hakka Territory. The River basin extends from Jiangxi Province and into Guangdong Province, and it is an important ecological resource in the region. Firstly, this part presents the background information of the region by providing an overview of the geography, history, and climate of the region. Water from the Dongjiang River, which serves the population in this Hakka territory and the first-tier cities in the Pearl River Delta is a key resource discussed in the following part. In the following sections other important regional features introduced are the local ethnicity, culture and also local tourism destinations. The Hakka local ethnicity, with the Capital of Hakka Culture in Meizhou, proves appealing to foreigners, and thus attracts many visitors to the region. The region has many tourist destinations contributing to the local economy, such as Changgang Town with tourism sites like the Taoist Mountain, Danxia Lake, Buddhist Temples, and Meizhou. Revenue from tourism in rural and poverty-stricken mountain communities in China is crucial for sustaining livelihoods. The Dongjiang River (DjR) (see Fig. 3.19) basin has three major supply reservoirs, i.e., Xinfengjiang, Fengshuba, and Baipenzhu, all which provide fresh water to the communities, supports agricultural and industrial activities in the Pearl River Delta (PRD). The hydrological system of Guangdong province is a crucial system of rivers and reservoirs as it supplies water to the Greater Bay Mega City (see Fig. 3.01); hence, an asset for urban development and environmental sustainability for the Global City. However, the quality and amount of water resources that are available in Southern China present a serious concern, especially due to increased economic development and urbanisation. Immense environmental pollution is a key challenge in the region. For example, high population growth, industrial and agricultural activities play a key role in the deterioration of the environment, and quality of water. Finally, this part discusses important national development policies such as rural spatial restructuring and land consolidation that are applicable to the rural and mountain territories of China and pertinent to addressing poverty and environmental degradation facing the region.

2.3.1. Regional Contextualization: Geography – History – Climate:

The hydrological system of the Dongjiang River Basin extends from Jiangxi Province and into Guangdong province and is particularly significant for the region (see Fig. 2.13). The territory, including its water transfer system linked to the River Basin (see Fig. 3.19), is a major water source for the Pearl River Delta. This source of water is critical, considering it serves metropolitan cities, as Hong Kong, Guangzhou, and Shenzhen (Yang, Chan, and Scheffran 2016). The purpose of this chapter is to provide background information and setting the scene for the design studies and scenario constructions in Part Three. This overview description of the Dongjiang River Basin and its surrounding Hakka territory is divided into nine subtopics. The first topic this part presents is its geography, which is particularly important for an understanding of the context of the Dongjiang River Basin and the Hakka ethnic territory in Guangdong. The next topic is history which offers a description of the Hakka history in the region. Information is provided on migration, language, customs, and architecture of the people in the region. The part later presents the local climate, including annual rainfall, precipitation, and characteristics of the seasons. Emphasis is put on how climate change and anthropogenic activities have an impact on the region. The next subtopic on the relevance of the territory due to the water supply to the Pearl River Delta shows that protection of water sources proves to be an important objective for the region. Environmental pollution and its impacts on the Dongjiang River basin is also discussed in this part as an urgent challenge the region faces. Other key assets of the territory include its construction material sector, tourism development & scenic destinations, its local Hakka culture and ethnicity. Finally, this part discusses national development policies applicable to the mountain territories in the region, noting the efforts and improvements that can effectively address the challenges that face the Dongjiang River basin & its surrounding Hakka territory (see Fig. 2.14).

Geography:

The Dongjiang River (DjR, also referred to as East River in English) is one of the three main branches of the Pearl River which is found in Southern China. The River has its starting point in Xunwu County in Jiangxi Province, and where it flow south-west to Guangdong Province. It then runs into the estuary of Pearl River of the South China Sea. The total length of the main stream of the DjR is 562 km, where 127 km of this length is in Jiangxi Province while the remaining part is in Guangdong Province (Yang, Chan, and Scheffran 2016). In other measures, the total area of the Dongjiang River Basin is 35,340 square kilometres, and the Guangdong portion accounts for ninety percent. The River Basin has three main reservoirs, namely Xinfengjiang, Baipenzhu, and Fengshuba. The three reservoirs are crucial in changing the river flows. It is important to point out that the water quality of the main stream is relatively good, even though the estuary or delta area and some tributaries have poor water quality.

The Geographic Context: Dongjiang River Basin in the Greater Bay cluster of cities.

Water resource Interdependency between the Mega-City and the Dongjiang River Basin.

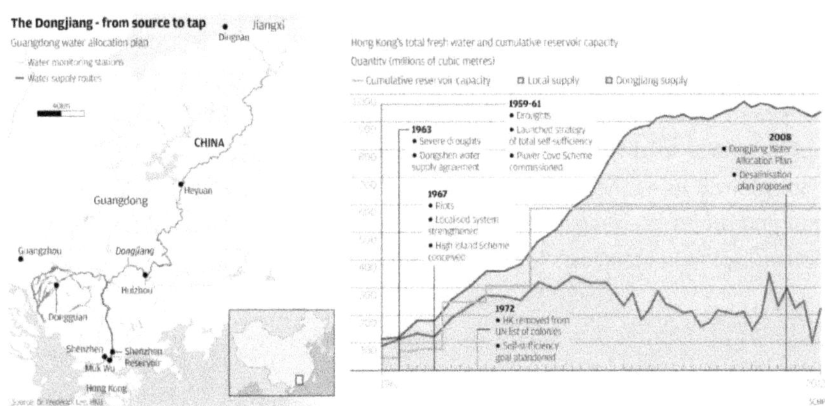

Fig. 2.13: Over-consumption of fresh water by wealthier cities of the Greater Bay cluster of cities causing future risks water shortage risks to the economically vulnerable population in the Dongjiang River Basin and the more developed cities in the territory. Additionally, environmental pollution, flood risks, potential future draughts undermine the resilience of the territory for unanticipated natural disasters. Source: SCMP, Ernest Kao, 2015.

The DjR basin plays a significant role in the territory. In particular, the River Basin is the main freshwater source for the region. In this light, the freshwater from DjR supports industrial and agricultural production and livelihood in the Pearl River Delta (PRD). According to Yang, Chan, and Scheffran (2016), the PRD had a total population of approx. 120 million in the year 2018, and this entails a combination of Global cities and industrial centres, i.e., Guangzhou, Hong Kong, and Shenzhen. The river management associated with the DjR carries other functions that include power generation, flood management, shipping, and also reducing pollutant discharge and salt intrusion. Yang, Chan, and Scheffran (2016) also note that recently, the DjR water management has faced numerous challenges such as urbanisation, population growth, reducing agricultural land use, industrial development, ecological degradation and fluctuation of river flow every year as a result of natural and man-induced factors such as climate change. This is a particularly important aspect when considering the resources and sustainability challenges of the Dongjiang River Basin and its dependent territories with communities living below subsistence level. The impoverished, peripheral areas outside of Meizhou, in the Upper DjR basin, are important mountain regions to consider due to its tourism potential. Its Hakka cultural & ethnic assets provide opportunities to diversify livelihood strategies for local communities. Meizhou, also referred to as the world's Hakka capital and a cultural city, is located at the merging point of Jiangxi, Guangdong and Fujian provinces in South-eastern China (Exploring Meizhou's Economic and Cultural Potential 2017).

History:

The Hakka Cultural Heritage is a significant contributor to economic diversification, thus a prior introduction of the history of Meizhou is pertinent to this study. Meizhou is found in the north-eastern part of Guangdong Province (see Fig.2.14). The city, which is a river town, sits on the northern bank of River Mei and it links it to the South China Sea, which is about 150 kilometres to the south through the Han River. Much of the eastern part of Meizhou, and the province itself, portrays a story of an ethnically different Guangdong compared to that found in Guangzhou and Shenzhen.

Internal migration of the Chinese, i.e., the Hakka people partly shapes the history of the Meizhou region and its mountainous hinterland. In the start of the fourth century, the Hakka people started moving from northern China to Guangdong and the areas around. With the migration, Hakka culture, dialect, and architecture were introduced into the region. Meizhou is a major centre of Hakka culture. Meizhou is also considered the capital of Hakka culture (Guangdong China,

2018). Hakka in Chinese translates to guest families (Kejia) but ironically, historians explain that Hakka people were not fundamentally perceived as welcoming guests. This was because competition for resources such as water, agricultural land, and sites for settlement increased in the 17th century. The Hakka people built strong, and unique structures/buildings meant to protect the town from attacks from the new neighbours, and at the same time, withstand natural disasters such as earthquakes (Pamhule: Around Meizhou, Guangdong 2016).The Upper Basin of the DjR (or Mei County) has six provincial industrial operations specializing in construction materials that help to sustain the economy of the region. These are Meizhou Guangdong Industrial Zone, Tangxia, Pingyuan, Xingning, and Guangzhou Haizhu. The others are Wuhua and Musa. Such regional knowledge is pertinent in understanding the industrial activities found in Mei County.

Climate:

The DjR Basin Territory lies in the subtropical monsoon climate zone and is located at low latitudes. Due to proximity to mountainous terrain, the South China Sea, and the Pacific regions, the territory experiences long summer, short winter, high annual temperature, hot and cold seasons, and high amount of sunshine. According to Chen, Zhang, Lu, Zhang, and Zhang (2011), several factors contribute to the increase in mean global temperature, and they include variability in climatic conditions and emission of greenhouse gases as a result of farming and industrial activities. Moreover, the Mei County region experiences abundance and concentration of rainwater and also flow occlusion (Zhou, Lai, Wang, Chen, Zeng, Chen and Bai 2018). Dongjiang River basin experiences a cool and dry period as well as a wet and warm period, with the two having approximately equal length (Sun, Xia, Xu, Guo, and Sun, 2016). As Zhou, et al. (2018) note the basin has mean annual precipitation of 1750 mm with a mean annual temperature of 21°Celsius. Climate change has significantly influenced water quantity and deterioration of water quality in the Dongjiang River Basin in South China (see Fig. 2.15) (Zhang, Xiao, Singh, and Chen 2013). Yang, Chan, and

Scheffran (2016) also note that the process of urbanization and climate change aggravate water shortages.

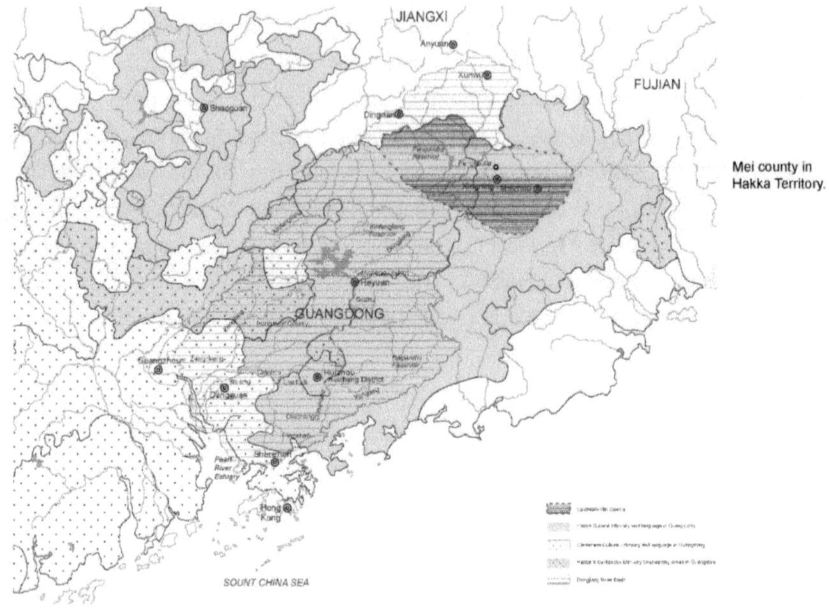

Fig. 2.14: Mei County in Guangdong province. The maps shows the hydrological system, indicative extent of Hakka and Cantonese ethnic distribution; and overlapping regions between the ethnic groups. For full Guangdong province map refer to Figure 1.02. Image generated by the author, 2019.

Partly, the Pearl River Delta Zone is in Guangdong Province, in the southern part of China, constitutes the downstream Lower River basin of the DjR territory. The region is in the subtropical climate zone. The zone has annual average precipitation and mean temperature per the year of 1600-2200 mm and 19.5-22.3°Celsius, respectively (Sun, Xia, Xu, Guo, and Sun, 2016). In particular, in the year 1998, the mean flow of local rivers in a year, or the mean runoff that was locally produced per year, was 42.05 billion m^3 whereas the per capita water resources was recorded at 1897 m^3 (Zhu et al. 2002). The amounts were lower compared to the average level, i.e., 2221 m^3 for China, and less than a quarter of the global level.

2.3.2. Importance of the DjR Territory due to Water Supply to the Pearl River Delta:

The supply reservoirs play an important role in environmental sustainability and future urban resilience. Shortage of water, therefore, slows down economic development, degrades ecology and also causes disputes and conflicts over water use sectors (He, Yang, and Chen 2018). The major causes of water shortage include water resources being unevenly available in limited places, the variability of this resource, and an unbalanced concentration of population centres in the region. The far-reaching implications of these factors are restricted economic activities and slow growth for peripheral areas in the territory.

The quality and amount of available water resources in Southern China is a serious concern. The underlying factors include fast rate of urbanisation and the overemphasis of economic development in only few cities of the region. As highlighted by Sun, Xia, Xu, Guo, and Sun (2016), the Dongjiang River is instrumental in southern China as it supplies water for irrigation and potable water to different parts of the Pearl River Delta (PRD). Therefore, the river delta in the south China region has experienced fast economic growth since the late 1970's due to drinking water provided from its remote hinterland.

The Dongjiang River is a main tributary of the Pearl River basin. In respect to annual average discharge and basin area, the river is the smallest tributary for the Pearl River basin. However, it is the fundamental source of a large population (Rong et al. 2018), who are settled in the PRD, one of the world's largest Mega Cities in South China (Freshwater Health Index 2017). Among the communities that depend on the river include Hong Kong that lies outside of the basin, but consumes the majority of its municipal raw water from the DjR. The greater provincial drainage area of the river flows through six urban regions in Guangdong Province.

The cities are Meizhou, and Heyuan located in the upstream, Guangzhou (Zengchen) and Huizhou in the middle stream, and Shenzhen and Dongguan which are in the downstream (Tan 2012; He, Yang, and Chen 2018). The Dongjiang River originate from Jiangxi Province and flows from northeast to southwest, and ultimately into the Pearl River estuary. The DjR is an essential source of water for the towns located downstream, i.e., Hong Kong and Shenzhen.

Over the past three decades, areas along the river greatly benefited as they have experienced economy growth and also social development (He, Yang, and Chen 2018). According to Tan (2012), the five towns of Guangdong Province, which are along the DjR basin are home to approximately 50% of the Province's population, and they further generate close to 70% of the total GDP of Guangdong. In addition, Hong Kong purchases approximately 80% of its fresh water from the Dongjiang. Theaters have turned out to be "water of politics," "water of life" and "water of economy" to both the DjR basin and Hong Kong (Tan 2012). For example, Hong Kong's domestic consumption has risen from approximately 190 litres every day per capita to 220 litres daily per capita from the year 1998 to 2015 (Blundy 2017).Tan (2012) notes that a weakened Dongjiang is a vulnerable Hong Kong as it is likely to destabilise the economy and livelihood of the residents in the town and also the economically dependent mountain areas in the upstream part of the basin. Therefore, the government could introduce urgent water tariff reforms to reduce water wastage (Blundy 2017).

The major towns that have experienced this benefit are Dongguan and Shenzhen in the downstream, and they account for the highest rate of water use as well as pollutants discharge into the river. On the other side, the settlements found upstream of the river around the Fengshuba reservoir, Xingning and the peripheral mountain areas of Meizhou are particularly disadvantaged as they face limited opportunities for livelihood diversification. Constrained economic development hinders the emancipation of rural settlements from the resource exploitation by metropolitan areas.

For instance, the individuals who live upwards of the DjR in the Fengshuba Reservoir valley are left with little other options than to supply scarce water resources to sustain Shenzhen and Hong Kong's consumption of water, instead of securing local supply of water for their community and economic development (Globalization Monitor Limited 2011). Eventually, this maintains quality and quantity of water resources in first-tier cities, but leads to a dispute between upstream and downstream, especially during the dry seasons (Liu et al. 2010). The ultimate challenge is environmental and ecological instability. Such challenges could be addressed through mechanisms for water conservation, recycling, and participatory allocation of resources (Yang, Chan, and Scheffran 2016; Shi, Chen, Wang, and Niu 2018).

2.3.3. Environmental Pollution:

Maintaining the quality of water from the DjR basin plays a particularly crucial role for the long-term resilience of the region. However, the contamination by industrial activities and pollution due to urban construction projects compromise the quality of water resources in Guangdong Province and other parts of China undergoing urbanization. Various environmental problems are evident in the region, and they cause degradation of the quality of water from the river, its supply basins and in ground water.

The environmental problems are the result of excessive waste generation that is discharged into the hydrological network of Guangdong. The Pearl River Delta continues to experience high population growth, and there is also increased pollution due to agricultural and industrial production. These factors threaten and also deteriorate the environment and river water quality (Liu et al. 2010). Sun, Xia,

Xu, Guo, and Sun (2016) note that despite this evidence, there is limited exploration of the pollution sources of the DjR.

In discussing the issue of environmental pollution, it is important to list the main types water contaminants found in the Dongjiang River. Major pollutants in the water environment include permanganate value (COD mn), dissolved oxygen (DO), biochemical oxygen demand, oils, and non-ionic ammonia (Zhu et al. 2002). Water pollution greatly threatens the ecology, environment, and human health of the rural settlements in which yet need to grow out of poverty. Rong et al. (2018) state that water pollution has been prominent in the past few decades due to the expanding export-based economy of the South East coastal region in China. Recent relevant research shows that continuous degradation of water quality is a significant driver that limits attaining sustainable growth of the economy and human capital formation (Zhu et al. 2002). Therefore, it is paramount to implement remedial measures that safeguard water environment for the next phases of urbanization in China.

According to the UN Sustainable Development Goals (UN SDG 2030), every individual has the right to access affordable and clean water, and which is adequate to guarantee fundamental living requirements with no discrimination. Similarly, Vollmer et al. (2018) hold that ensuring freshwater security presents a major natural resource challenge. While the DjR hydrological network provides water to the Pearl River Delta, there is uncontrolled pollution and excessive use of water resources by modern agricultural activities and highly developed industries (Zhou, Lai, Wang, Chen, Zeng, Chen and Bai 2018). Industrialising areas such as the coastal region of Guangdong Province increasingly face severe water pollution. According to Globalization Monitor Limited (2011) report, half of China's groundwater is polluted, causing the emergence of "cancer villages" in rural and mountain communities. Such villages threaten the health and livelihoods of local residents.

Types of contaminants based on water sampling points:

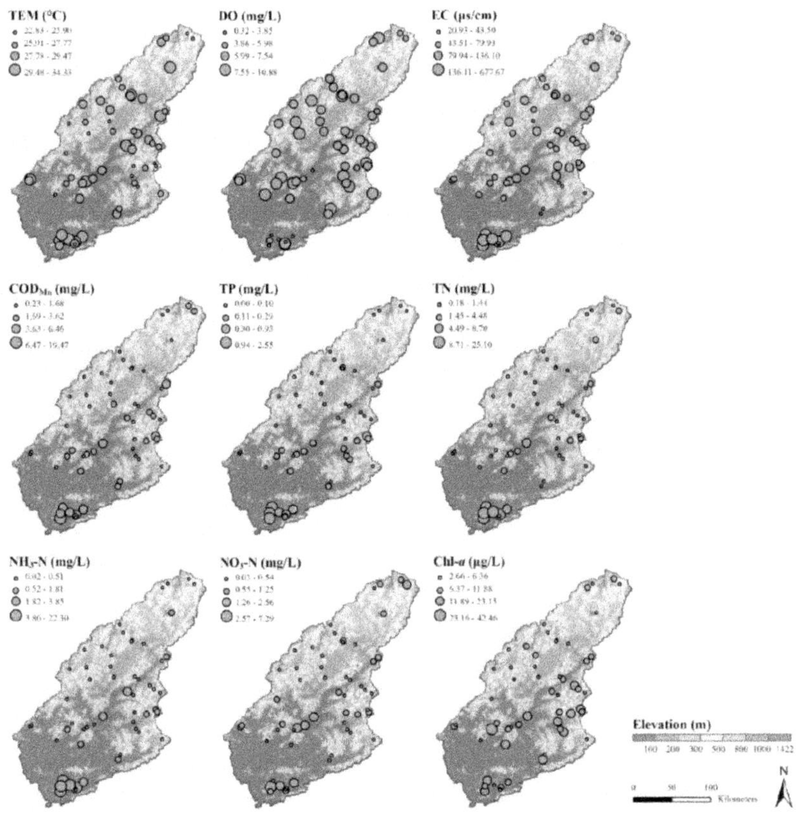

Fig. 2.15: Water contamination data in the territory. Image Source: Ding, Jiang, Liua, Hou, Liao, Fud, Peng, 2016.

Water Scarcity is a serious environmental issue, undermining the growth and resilience of the Pearl River Delta and its hinterland.

Fig. 2.16: The water reservoirs are a key ecological resource for the Greater Bay territory. Source: Evergreen, SCMP 2011.

The cause of the increased upstream and downstream water use disputes, between the disadvantaged settlements and the cities in the delta regions, is that water resources are largely used for revenue generation where large-scale agricultural and industrial firms disregard paying for social and environmental damages caused by resource extraction (Globalization Monitor Limited 2011). The negative outcome is then water injustice. The irony is that most of the rural inhabitants in the water supply valleys are not the beneficiaries of the industrial development or revenue from the water supply.

2.3.4. Local Economies:

Due to the concentration of poverty, an urgent area of consideration is the Upper Basin zone in Mei County. This territory includes the Fengshuba Reservoir, Heshui Reservoir and the peripheral areas along Mei River. It is located at the junction of Jiangxi, Guangdong, and Fujian provinces. The Mei County is furthest away from the affluent PRD zone, yet it is an essential territory that links coastal cities in Guangdong and the landlocked mountain regions. It is hoped, it will transforms into an important future transport and economic hub. Various features characterise Mei Country, and they include an appealing natural water reservoir scenery (see Fig. 2.17) and its Hakka cultural ambience. Besides, in recent years, Mei County has undergone promising economic growth due to the tourism sector and construction material industry. For example, the first three-quarters of 2016 proved promising for the region's GDP reached 7.1%, i.e., is 0.4% higher than the national average (Exploring Meizhou's Economic and Cultural Potential 2017).

Water reservoirs as an ecological resource and scenic landscape attracting visitors from cities predominantly Guangdong province. Conference centers and a golf course are nestled in the picturesque setting outside of Fengshuba reservoir.

Fig. 2.17: The Fengshuba reservoir was visited during field trip visits. The region can be reached in one hour by car from Meizhou which is one hour flight away from the Guangzhou Megacity. Photography: by the author.

Tea plantations as manicured landscape parks and agricultural tourism destinations.

Fig. 2.18: Tea is an important export product and Yannafei tea plantations as parks frequented by holiday makers. Photography: by the author.

Pottery, ceramics and handmade wood furniture and artifact are being offers as goods as local products.

Fig. 2.19: Picture taken in a tea house in the Yannanfei Hakka tea plantations outside of Meizhou, in Mei Xian. Photography: by the author.

In the Mei County mountain region, farming villages from the Maoist period are open to visitors as cultural heritage. This particular stone house is in Shizhai village. Temples in the mountains are being visited from tourists mostly coming from within the Guangdong province.

Fig. 2.20: Villages and Mountain Temples were visited during field trips guided by a local tour guide. Photography: by the author.

Village house from the Maoist period as a visitors centre preserving interiors from the 1960s and small exhibition of Chinese artifacts displayed as cultural heritage.

Fig. 2.21: Interior views of courtyard and the living room in visitors centre. Photography: by the author.

As a convenient transport channel and richness in Hakka cultural tourism (see Fig. 2.18-2.21), increasingly Meizhou plays a role in the poverty alleviation of its mountain region. Meizhou is expanding a regional transport network with a small airport and by accelerating the construction of a high-speed railway. The Meizhou-Shantou High-speed Railway (see Fig. 2.22) is targeted to be upgraded by the end of the year 2019 (Exploring Meizhou's Economic and Cultural Potential 2017). The completed expressway is expected make travelling efficient, and attract more visitors in the future.

The Dongjiang basin has undergone incremental societal and economic transformation over the past three decades. The Gross Domestic Product and population of regions such as Huizhou, Dongguan, and Heyuan in the Midstream zone that surround the river basin have kept a persistent and rate of economic development as a result of China's opening policy and reforms since 1978. In particular, the GDP of Dongguan increased from 8 billion Chinese Yuan in the year 1980 to 262.7 billion Chinese Yuan in the year 2006 and had an average growth rate of 14.34% (He, Lin and Chen 2013). In particular the population growth and changes in land use have been critical to the economic development in the region.

Moreover, Lee and Moss (2014) note that the agricultural sector dominates the economies of Xunwu and Heyuan, which are Middle basin zone. The cities in the middle and lower catchments, i.e., Dongguan, Shenzhen, and Huizhou highly concentrate on industrial production since the 1990s. Nonetheless, the structures of both Shenzhen and Dongguan have changed in the past five years as they currently favour the tertiary sector. The tertiary sector dominates the economies of both Hong Kong and Guangzhou, regions located at the western and southern edges of the Dongjiang River basin, respectively (see Fig. 3.19). According to Lee and Moss (2014), the populations that depend on water from the DjR include those in three impoverished rural counties namely Xunwu, Dingnan, and Anyuan in southern Jiangxi, which are made up of subsistence-level income earners.

2.3.5. Local Tourism Destinations and Development:

In the last three decades, tourism activities in China have significantly improved, for remote mountain communities. Rural tourism in the country proves a potential catalyst that is effective in promoting agricultural development, industrial restructuring, improvement or upgrading of reserve areas, and livelihood diversification (Wang, Cheng, Zhong, Mu, Dhruba, and Ren 2013). For instance, China's rural areas are increasingly being challenged by issues of reducing welfare service provision, agricultural restructuring, depopulation, degradation of the natural environment, and also deficits in infrastructural amenities and limited adaptation to renewable sources of energy (Wang, Cheng, Zhong, Mu, Dhruba, and Ren 2013).

China's rail infrastructural network with new High-Speed rail connections or upgrades of conventional rail to high-speed trains. In the future faster rail connectivity into the Inner provinces may support higher tourism turnover in remote territories.

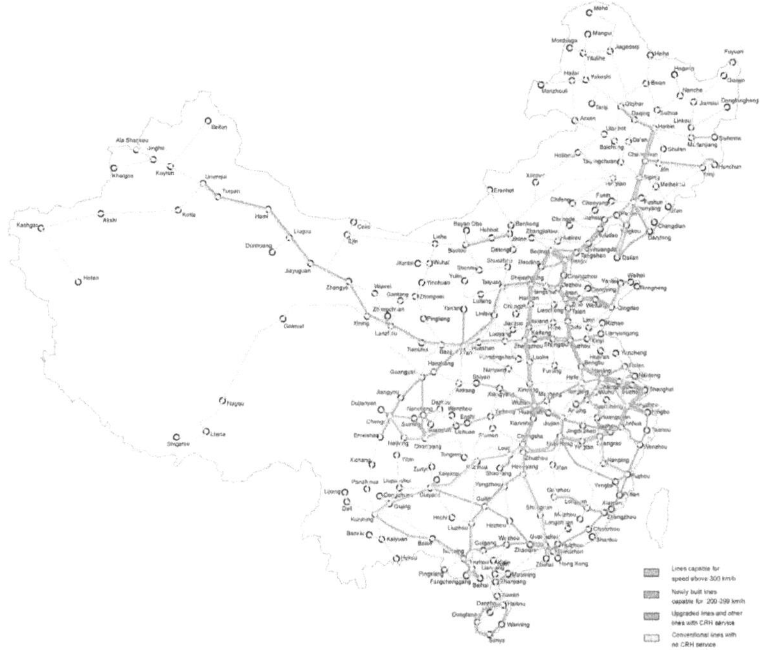

Fig. 2.22: Railway Infrastructural Integration in China: A new connection from Hong Kong West Kowloon Station, the Express Rail Link (XRL) is linked with the countrywide high-speed rail network in Greater China. In total the rail network is 25,000km long and constitutes of eight vertical and eight horizontal lines integrating the larger urban centers in China. Sources: Howchou & Express Rail Link (XRL), Transport and Housing Bureau.

Tourism development is regarded as promising. Similarly, tourism development is a significant driving force for poverty alleviation, protection of the scenic environment and ultimately, sustainable development. As Perales (2002) notes, since the 1970s, tourism in rural regions has increased significantly in all the first world nations in the world, and this is crucial to the development of the regions not only from an economic but also environmental point of view. For a better understanding of the opportunities of tourism development, it is pertinent to distinguish between different types of rural tourism. According to Zhang and Chen (2003), the different types include rural leisure tourism, 'agritainment' tourism, rural travel, rural eco-tourism, farm tourism, and also rural customs and ethnic tourism. However, the concept of rural tourism, as well as the different forms of agricultural tourism offerings, are gradually transforming with the development of cultural ethnic tourism in China (see Fig. 2.25). It is for this reason, that rural regions hold activities that promote local culture and farming tourism, with the understanding that it helps to maintain local heritage and existing food industries in the regions.

The government of China plays an important role in supporting the development of rural tourism. Wang, Cheng, Zhong, Mu, Dhruba, and Ren (2013) emphasise that all local administrations place a great importance to rural tourism in China. In particular, the government acknowledges rural tourism as a means to assist development and rural income diversification. There are various ways in which the government shows support for the development of rural tourism. First, there are numerous development programs and legislation that local, provincial, and central government has created which promote an atmosphere as well as policy-orientation necessary to develop rural tourism all over the country. In other words, the Chinese government ensures that there are necessary or appropriate regulations and laws pertinent to the realisation of standardised provincial management of rural tourism in different provinces of the country. In terms of policy, the Chinese government enacted the National Rural Tourism Development Programme (2009) in the year 2009. With this programme, it was evident that the concept of developing rural tourism was part of China's national development plan, and as such, implemented on a national scale (Wang, Cheng, Zhong, Mu, Dhruba, and Ren 2013). In addition to the role government departments play in promoting rural tourism, farmers and residents in mountainous communities of

China have an important role to play in attracting more urbanites to visit the countryside for vacations (Caixiong 2018).

Furthermore, the setting up of specialised agencies of managing rural tourism ensures effectiveness in efforts such as planning, advertising and supervision of rural tourism (see Fig.2.24). Similarly, tourism fairs significantly promote tourism in different rural regions in the country. For instance, Guangdong has held multiple fairs that focus on rural tourism (Caixiong 2018). As an example Caixiong (2018) notes a special tourism exchange and fair that was held in 2008 in Yunfu, a city located in the mountains of western Guangdong province. The tourism fair was prepared to boost rural tourism and economy (Caixiong 2018). During the fair, Yuan Huoyue, the head of the Xinxing county government praised rural regions in the mountainous regions of western Guangdong for their suitability for tourism. The regions have characteristic fresh air as well as beautiful scenery, both of which are excellent features and potential for the growth of rural tourism. As the head of the county government noted, development of rural tourism in the Guangdong region is pertinent in boosting the rural economy of the region.

Moreover, the National Development and Reform Commission, China's Ministry of Culture and Tourism and 15 other government departments jointly released a guideline on how to sustainably develop rural tourism (Xinhua Net 2018). The guideline has five key measures such as improving public service and infrastructure for convenience in rural tourism and also strengthening planning for better coordination of regional development in rural tourism (Xinhua Net 2018). The other measures entail improving quality of products and services, focusing on the benefits of rural tourism in alleviating poverty among rural residents and also improving innovation in rural tourism marketing.

Such efforts by the state indicate that government-led rural tourism growth in China aims for healthy and sustainable rural tourism development, while trying to address poverty in the countryside. Considering the one-party political system in China, in an ideal case, governments at all levels facilitate the growth of tourism in the rural areas which preserves authentic ethnic cultures and its environment. Among the important roles the state plays include preservation of historic buildings, maintaining of local traditions and dialects, public promotion and marketing, macro-guidance, financial support and policy guidance (Wang, Cheng, Zhong, Mu, Dhruba, and Ren 2013). Nonetheless, for a sustainable development of tourism in rural regions of China, the government could collaborate with other stakeholders such as small scale venue operators and private developers, local residents and tourists themselves. Wang, Cheng, Zhong, Mu, Dhruba, and Ren (2013) propose that for sustainable development, it is highly important to bring on board all participating stakeholders and disciplines.

The Development of the Tourist industry in Rural China is promoted as a way to alleviate rural poverty and to diversify livelihood strategies.

Fig. 2.23: Tourism in the Chinese countryside is targeted at Chinese urbanites and visitors from abroad. Source: National Geographic Chinese Edition, 2016, TimeOut Shanghai, 2016.

Commercialization of Ethnic and Agricultural tourism in an experience economy.

Fig. 2.24: Thematic categorization of Cultural, farming and culinary tourist experience can be selected on the internet and travel agents. Various sources: Time- Out Shanghai, 2016.

Tourism Potentials:

In particular, cultural tourism the Guangdong Hakka ethnic region helps to diversify incomes and in the quest for a 'harmonious society' of ethnic groups that were previously underestimated in impoverished rural and mountain regions. Guangdong Province is located in the southern most edge of the Chinese mainland and it neighbours the South China Sea in the south and the northern mountains territories. The area is characterised by beautiful natural landscape features and ecological resources as well as a historical cultural importance (Hui, Jiemin, He, and Qi 2018). The large-scale regional advantages drive reform a relative proximity to highly developed metropolitan areas with international export harbours. Besides, having transport linkages to Macao and Hong Kong helps in the economic growth of the province, and the growing tourism sector. According to Hui, Jiemin, He, and Qi (2018), the tourism industry continues to rank first in China, thus making Guangdong rank first in the country for five consecutive years. This industries revenue rose eightfold, i.e., from 114.995 billion yuan in the year 2000 to over 908 billion yuan in 2015. The average growth per year was 57.57 billion yuan, apart from the limited growth in 2002 and 2003. Yue (2018) also emphasise the essence of tourism to local governments, fundamentally noting that the governments remain committed to reduction of poverty in mountain villages. Despite the global economic slowdown, China's rural tourism market has been continuously booming, considering 1.36 billion tourists toured China's rural areas in 2016. Researchers estimated that the visitors spent 400 billion yuan on the trips made (Yue 2018). With such contributions, local tourism in mountainous regions of China plays a crucial role in poverty reduction. ChinaDaily.com (2019) promotes the idea that the booming rural tourism benefits rural residents, thus assisting in lifting the villagers out of poverty.

However, despite the commendable achievements in Guangdong Province, the development of tourism in the region faces severe problems. Hui, Jiemin, He, and Qi (2018) cite slow pace in development, which ultimately make the overall advantages less obvious in comparison to other main domestic tourism centres. For instance, between the year 2010 and 2015, the yearly average growth rate of tourism revenue in Guangdong province stood at 19.9%, which was below the national level by 2.1% over a similar period. While the total revenue from tourism was ranked first in the year 2016, other provinces such as Zhejiang and Jiangsu were rapidly catching up. Moreover, Guangdong experience a lag in visitor arrivals, compared to Jiangsu and Zhejiang in the same year. Another key challenge is a prominent imbalance in tourism growth between different prefecture-level towns in Guangdong (Hui, Jiemin, He, and Qi 2018). Hui, Jiemin, He, and Qi (2018) further note that the traditional advantages of Hong Kong as a preferred window of entry and exit into Guangdong by tourists, and as a frontier of China's opening up and reform are increasingly diminishing. Considering the emerging competitive pattern of the Chinese tourism industry, it remains pertinent to develop ways to improve tourist related factors. Such factors include safety, political stability, market demand, tourism resources, reliability of transport infrastructure, and traffic location for purposes of reconstructing the competitive advantage of Guangdong in tourism development (Hui, Jiemin, He, and Qi 2018).

Guangdong is a major tourists' destination, and with government support and contribution, the province could continue to attract an increased number of tourists. For instance, Xinxing county government remains committed to developing rural tourism. The targeted tourist sites include Zen Buddhist culture, rural scenery, ecology, and leisure, and also hot springs. Xinxing is the hometown of Huineng (638-713) and is the sixth and last Patriarch related to Chan Buddhism. Besides, Huineng is an influential figure found in Japanese Zen Buddhism, and it originated from Chan Buddhism. There are three Oriental saints Laozi, Confucius and Huineng; hence, the benefits of the latter cannot be overlooked. Therefore, the effort to improve the tourist sites in Guangdong will go a long way in re-doubling development of rural tourism.

Changgang Town is also a suitable tourist destination. The town is located in the northern region of Xingguo County and is rich in natural resources (Yao and Xie 2016). For example, the water resources in Changgang town entail the national large-scale Changgang reservoir as well as eight other reservoirs that are small but economically important in the region. In addition, the town has tourist resources such as the famous Taoist Mountain, Danxia Lake; Red tourist attraction sites, i.e., the Memorial of Chairman Mao at Changgang and also various Buddhist temples (Yao and Xie 2016).

Another key tourist destination is Meizhou in Mei County, which is a city in the East of Guangdong province. There are numerous things tourists can do and see in Meizhou, most of which are Hakka culture related. For example, the town has numerous ancient houses that are nestled among newer buildings, with such structures showing genuine traditional architectural settings that give the city its identity as well as maintain the ethnic heritage as opposed to an international city serving the global economy. There are numerous tourist destination sites or attractions in Meizhou such as traditional houses (see Fig. 2.26) which are ancient Hakka homes and also the Hakka museum. The Museum is the largest Hakka museum in the country, with the main building consisting of five halls i.e. a space for the origin of Hakka, customs, architecture, culture and the genealogy of Hakka. The Lianfang Tower and Yinna Mountain are also important tourist destinations in the region.

Hakka Traditional Villages and Preservation of Local Ethnic Culture.

Fig. 2.25: Walled Hakka Villages are popular tourist attractions. Source: Jimmy Ng.

Hakka Architectural Heritage - Southeast China Toulou Earth Buildings as 'Small Villages'.

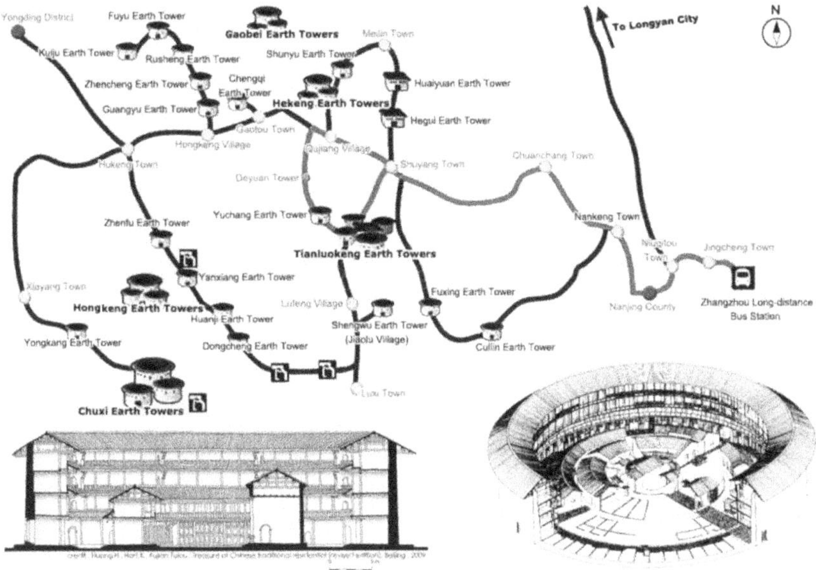

Fig. 2.25a: Tulous round buildings are an important cultural asset of the Hakka Diaspora in Southeast China. Many of the earth towers were built between the 13th and 20th centuries. Today guided architectural tours provide access for visitors into the fortified Hakka villages. Tulous can house up to 600 people. Some of the Tulous offer accommodation for tourist to experience the unique Hakka history and culture. Source: Huang et Hart; Ronald Knapp; Travelchinaguide.

Examples of Hakka Culinary Traditions as opportunities for Ethnic-cultural food tourism.

Fig. 2.26: Hakka cuisine restaurant and menu visited during field trips. The restaurant in Chiguang Town specializes in traditional Hakka recipes. Source: by the author.

2.3.6. Local Ethnicity and Hakka Cultural background:

The local ethnic culture is a key asset for the tourism development of disadvantaged rural communities. Thus, an explanation of its origins and background is provided in this part. Hakka represents a nomadic section of the Han ethnic group (Meizhou, capital of Hakka culture - Guangdong, China 2018). The Hakka group lived in northern China but later migrated south, escaping upheavals and unrest caused by famine and war during the start of the Jin Dynasty (AD 265-420). While the group lives with individuals from different cultures, they communicate through a traditional Hakka language, which phonetically is the same as Cantonese (Meizhou, capital of Hakka culture - Guangdong, China 2018).

The Hakka culture is the primary culture of Meizhou, and it is preserved in various forms. Some of the forms include the rural dwellings, i.e., round-dragon houses, soil of fence, and Zouma. The history of Hakka round house is particularly rich regarding architectural style, and many experts consider it the oriental version of the panopticon building typology. Hakka traditions are found with special local products, for instance, Meizhou Shatian shaddock and cannonball taro. As a Han Chinese people, the Hakka clan is considered to have started from a Song dynasty. The Mei County is also famous for its ballads, which are believed to have emerged during the Song Dynasty, i.e., 960-1279 AD. The local Hakka ballads are a combination of Hakka and other minority cultures, and its lyrics resemble those of ancient Chinese poetry. Most of the ballads are passed down to generations through oral history, mountain songs and they tend to touch on the topic of love.

While Mei County is not well-known for its food, there is a variety of Hakka cuisine, and it has different well-known staples (see Fig. 2.27). Among them are salt-baked chicken, chicken claws, chicken tripe, braised pork with preserved (dried) vegetables, Hakka meatballs, Meixian region maltose, Changle shochu Wuhua - Hakka meat stuffed tofu, preserved beef, candied ginger and wine of Hakka. The region is also famous for its Dancong Oolong tea, which is a group of thin-strip style oolong teas sourced from Guangdong Province (Meizhou, capital of Hakka culture - Guangdong, China 2018). Over generations, the Hakka style has remained relatively unchanged despite the transformations in Chinese society since the late 1970's reforms in the country. The Mei County also has a rich Hakka heritage characterised by unique and rich folk. The territory is rich in cultural heritage and is also surrounded by picturesque landscapes and water reservoirs (Exploring Meizhou's Economic and Cultural Potential 2017). As emphasized by Wang, Cheng, Zhong, Mu, Dhruba, and Ren (2013) designing a rural tourism development strategy requires full respect and understanding of the local customs, social norms, as well as maintaining its traditional culture, and the unique features for the revealing of special tourism destinations.

Notably, the Hakka ethnic heritage, philosophical, political and cultural legacy attracts visitors also to History tourism sites not only in Mei County, but also beyond provincial boundaries in China. Examples of the Hakka diasporas (see Fig.2.26a) are mentioned for reference in the last part of this section. The influence of this ethnic group on collective housing (see Fig.2.26c), architecture and community planning continues to be a paradigm visited by planning professionals and the general public.

Hakka Cultural Diaspora and Sun Yat-Sen Heritage as a significant cultural capital supporting tourism:

Ethnic characteristics, independent - mindedness and its unique influence on the history and culture in China.

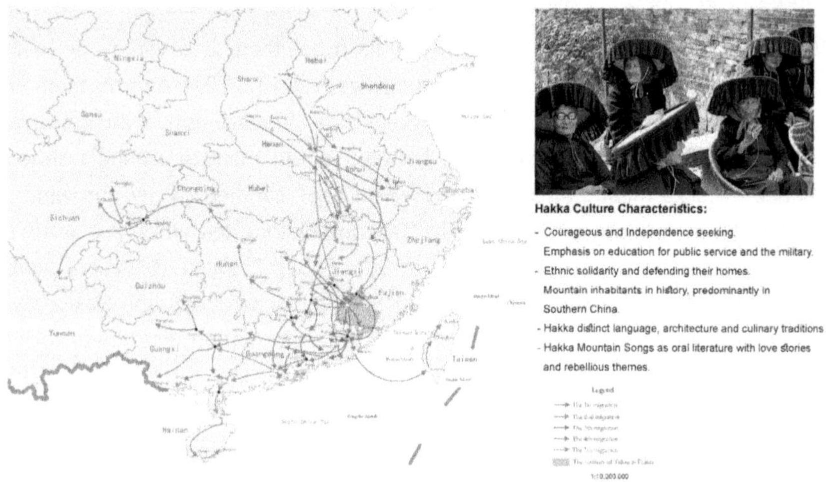

Fig. 2.26a: The Hakka ethnic group from Guangzhou had five major phases of migration in history: Phase 1 317-819; Phase 2 880-1126; Phase 3 1127-1624; Phase 4 1645-1867; and Phase5 from 1867. Today Hakka people are dispersed all over the world and continue to form distinct immigrant communities outside of China. Source: Lei Hua, Chongcheng Chen,Hui Fang, Xiangxiang Wanga, AsiaSociety.

Tourists tracing the vestiges of Hakka Revolutionary Leader: Dr. Sun Yat-sen (1866-1925), Founding Father of Modern China.

"History Travel" Destinations including museums, memorials, parks for Dr. Sun Yat-sen have been set up by the tourism boards in Guangdong Province, Hong Kong and also Jiangsu Province.

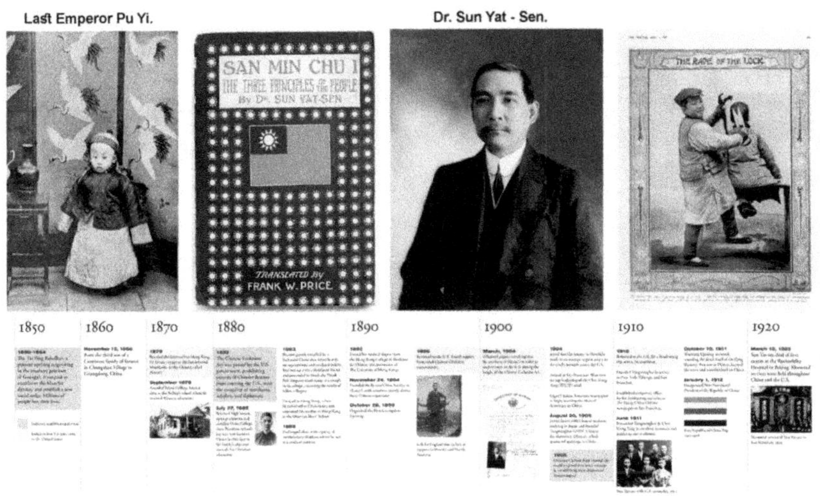

Fig. 2.26b: An independent-minded ethnic group: Hakka culture emphasizes education and rebellious temperament. Sun Yat-Sen was one of the most revolutionary politician who is known as the "the father of modern China". He became the first President of the Republic of China and founder of the Kuomintang (Nationalist Party of China).The above time line shows Sun Yat-Sen's life recalling his revolutionary and political struggles. Sources: Repro-tableaux, Hsu Chung-mao for ThinkChina, CHSA Museum. https://chsa.org/exhibits/online-exhibits/sun-yat-sen-an-american-legacy/life-and-times-of-dr-sun-yat-sen/ ; (for reference also see Dr. Sun Yat-sen Historical-Trail in: https://www.discoverhongkong.com/in/explore/culture/dr-sun-yat-sen-historical-trail.html).

Hakka Culture and Collective Housing traditions: Tulou building typology as a paradigm for community living continues to be a reference for collective housing projects in China.

Fig. 2.26c: Community building: Hakka collective housing examples and ideal communities in the countryside continue to influence settlement planning and housing projects in China.

Hakka culture and Sun Yat-Sen Heritage tourism example:

Sun Yat-sen Mausoleum on Zijin Shan - Purple Mountain, Jiangsu Province.

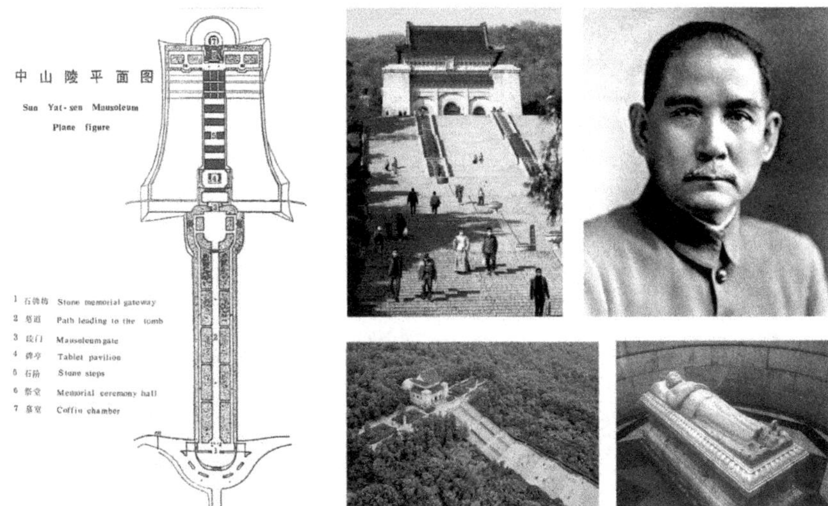

Fig. 2.26d: Dr. Sun Yat-Sen has left traces in society, politics and culture inside and outside of China. Several monuments commemorating Sun Yat-Sen as the 'father of modern China' are popular tourist destinations attracting visitor from all over the world. Dr. Sun Yat Sen's legacy continues to be a cultural asset for promoting ethnic-cultural tourism in China. Sources: China Heritage, Lu Yanzhi Architect, Yao Qian and Gu Bing ,Yang Bo China News Service, Portrait photo author unknown, Jiangsu Province Radio and Television Station, Coffin photograph by Lothar Ledderose.

Moreover, the prominent Hakka revolutionary leader, Sun Yat-Sen, who became the first provisional president of the Republic of China, following the 1911 Xinhai Revolution, aimed at ending Imperial rule, and to overthrow the last Emperor Puyi, who at the time was six years old (see Fig.2.26b) attracts visitors to historic sites and monuments . Sun Yat-sen's legacy, his influential political philosophy are the: Three Principles of the People' (1929): 1) Mínzú Zhǔyì or Civic Nationalism. 2) Mínquán Zhǔyì or Governance Rights. 3) Mínshēng Zhǔyì or Welfare Rights. A fuller account of Hakka influence in China and Dr. Sun Yat-sen's impact on the Hakka ethnic regions are substantial and therefore would need to be treated further in a separate study.

Fig. 2.26e: Chinese ink landscape painting shows the Sun Yat-sen Mausoleum on Zijin Shan - Purple Mountain. Source: Fu Boashi, 1962.

2.3.7. National Development Policies Applicable to Rural and Mountain Territories in China:

Rural China experiences dramatic transformations in how communities relate to the environment. The primary reasons for such transformations are urbanisation and industrialisation. Given that both of the factors are increasing, there is evidence of persistent widening of construction land, causing displacement, further fragmentation and diminishing of farmlands that were initially highly productive (Yao and Xie, 2016). Ultimately, this threatens national food security.

Further, development in rural and mountainous territories affects vital portions of ecologically sensitive land. Due to the steeper topography, difficulty of access and cultivation of land, mountain regions are less considered for national development strategies (Cuo, 2010). Any development targeting such areas tend to have a lower priority is often not considered holistically. For instance, according to Yao and Xie (2016), development occurs in the form of uncoordinated and scattered expansion of constructed land, overexploitation of fertile soil, and conversion of arable land for the purpose of development. This generates a patchwork of industrial, farming and construction uses of land. Complex repercussions of such development include environmental destruction, reduced biodiversity, and also the erosion of cultivable soil. In this light, there is deterioration of the rural environment, and weakening of ecological protection and eco-system functions, which ultimately threatens the security of the regional ecology (Yao and Xie, 2016). This further negatively impacts sustainability of both urban areas and rural mountainous regions. Integrated development in the mountain and rural regions in China is recorded to be lagging. Fundamental causes are associated with social, economic, geographical, and political challenges. Cuo (2010) states that approx. 70% of China's land area is categorised as mountain regions, and it is where approximately 26% of the total rural population resides. It is undeniable that rapid economic growth due to informatization, and the access to state-of-the-art knowledge has lead to poverty reduction in China in the past thirty years. Yet, the most acute rural poverty still remains concentrated in mountainous areas and requires more policy attention. Therefore, it is crucial to establish national development policies for the restructuring mountain regions in China.

One such policy could be rural restructuring in China. Rural spatial restructuring could help in establishing effective and intensive spaces for industrial activity, food production, environmental preservation and community settlement. Besides, spaces under rural restructuring of modest, yet functional habitats and are also ecological habitat areas (Long, 2014). Rural spatial restructuring can be considered a consequential evolution of the rural mountain areas driven by the economic and societal transformations. A new type urbanization phase in China should entail the emancipation of mountain areas from mega-city clusters, avoid dependencies on metropolitan areas, and support a self-renewal process of rural regions (Yao and Xie, 2016).

In an experimental study, for instance Yao and Xie (2016) examine, at a village-township level, the policy of rural spatial restructuring in an ecologically fragile mountainous area and rural region of southern China. According to their study, there are various methods for rural spatial restructuring based on landscape security pattern construction, and a model of expanding rural residential land by the use of scenario simulation and optimisation methods (Yao and Xie, 2016). One of the feasible policies in the ecologically fragile mountainous area is use of the security pattern under a necessary scenario. The scenario is pertinent as it can fully satisfy land requirements of agricultural production, food independence, ecological security, and socioeconomic development of the region.

There is limited research on rural spatial restructuring interested in scenario alternatives for China's rural development, especially studies focusing on optimisation of rural production, welfare and social security systems, environmental remediation, and biodiversity habitats at the village-township level. Therefore, it becomes challenging but also an exciting attempt to understand how rural spatial restructuring could impact ecologically fragile rural and mountainous regions. Chen and Xie (2016) note that rural residential land is found in small and scattered patches in most areas of China, particularly in the mountains. Therefore, it is necessary to optimise land use at a township or village scale, as studies focusing only on provincial scale maps overlook small and scattered settlements in mountain areas that are often hidden in surrounding features of other land use types (Yao and Xiem 2016). Thus, such studies make it difficult to establish an effective way to optimise spaces for dwelling, spaces for welfare provision, rural production, and sites for environmental remediation.

Extreme regional discrepancies are also common in both socioeconomic development, and biophysical and geographical conditions in China, considering the country is overly vast. Besides, there are differences in rural land-use conflicts in various types of regions, i.e. the problem of transformation of high-quality farming land to construction land in plains regions (Yao and Xie, 2016). On the other side, the conflict in the ecologically fragile mountainous areas relates to the conversion and fragmentation of ecological land by construction and development activities (Long and Liu, 2016). Exploration of various restructuring paths following local conditions identifies land consolidation as a potential national development policy in mountainous regions.

As a development policy, land consolidation is a necessary form of spatial restructuring in mountainous areas in China. This implies, to restructure rural production, ecological and residential space through consolidating scattered settlements in rural regions. According to Long (2014), consolidating fragmented pieces of land helps establish a platform for constructing new countryside and realise an alternative path for urbanity outside of mega-cities. Concepts of land consolidation and rural restructuring have attracted interest in academic circles in the world. Consolidation is a land management measure for spatial problem-solving. This intervention attempts to reduce different types of land fragmentation, increase productivity of rural land, and improve rural production and diversification of livelihood strategies for an independent rural development. The outcome is attained through a process of bringing plots together or rejuvenating derelict and disadvantaged rural settlements, mining and industrial land that have been abandoned. Such restructuring should be accompanied by the development of irrigation facilities, new roads, and provision of social welfare services and renewable sources of energy (Huang, Li, Chen, and Li,2010).

In China, the Ministry of Land and Resources established the Land Consolidation and Rehabilitation Centre in 1998 with some land consolidation projects initiated at the national level. The objective of land consolidation in China has changed over time. For instance, at first, i.e. in 1998, the national function of land consolidation was to increase the quantity of farm land through reclamation and consolidation. Another objective was to compensate the decrease of fertile land due to increased industrialization and urbanisation. Recently, land consolidation has transformed into full consolidation of water, agricultural, roads, settlements, and forestry, particularly those focusing on constructing collectivised agricultural land of high standard. In particular, between 2000 and 2010, land consolidation enabled the supply of 2.8 million ha farmland, and building of 13.3 million ha farmland with high level and stable productivity (Long, 2014). Such consolidation of vast rural areas in China proves an indispensable national development approach, particularly for regional food independence and promoting the self-sufficiency of mountainous settlements in the country.

Various types of land consolidation are applicable in rural and mountainous areas in China. Such measures may support long-term independent development in the mountains. Long (2014) details different types of consolidations that facilitate an overhaul of rural productivity, ecological and living space. The first type of agricultural land consolidation is aimed at increasing the quality and area of farmland and improving conditions for agricultural efficiency. Agricultural land consolidation includes the upgrading of drainage and irrigation, land leveling, paving field roads, protecting farmland, conserving the ecological environment, and other measures needed to cultivate the land. The aim is to enable the construction of farmland infrastructure, optimize the layout of farmland designed for multi-functional land use, maintaining high-standard farmland, and effectively facilitate the linking and concentrating of farmland plots so that farmers can engage in large-scale management (Long, 2014). The result could lead to agricultural modernization and an alternative urbanity in China's mountain regions.

Another type is mining and industrial land consolidation. In essence, this policy remediates industrial land of the township and village enterprises (TVEs) (Long, 2014). Similarly, the policy reuses mining land, i.e., ore, quarry, salt, and sand mining by considering new multiple functions for such sites. Additional functions of industrial and mining land consolidation include enhancing land saving, improving the value of such land, considering the potential rural-urban growth space, and introducing human settlement areas. The focus of such consolidation includes environmental remediation of rural land that has been deteriorated due industrial pollution. The objective is the efficient of use of land, and modernizing rural production, preserving inhabitability of settlement spaces and the ecological environment.

A further measure relevant for mountain areas, is the hallowed villages consolidation. This form of consolidation refers to rejuvenating disperse, idle, abandoned, and low-efficiently used construction land in the countryside, particularly rural housing land. The purpose is to improve public service facilities, rural mobility infrastructure, provision of a welfare services, improve living conditions and rural production. Additionally, it entails the re-use of construction land in the rural region to reduce the loss of farm land (Sun, Liu, and Xiu, 2011). Besides, hallowed villages consolidation is relevant in promoting the creation of new countryside, and finally, in achieving a new urbanity in the mountains. Long (2014) asserts that the fundamental goal of this approach is to improve social welfare facilities for communities and rural infrastructures through the recycling of derelict rural building land and neglected mountain farms.

In the Research by design Part of this book, case studies have been carried out in three mountain villages: Chiguang town, Longbei town, and Changsha town in Mei County. The transect as a planning tool has been tested for the mountain village sites. Adaptations to the transect tool have been suggested to best propose design and planning solutions in the distressed and impoverished village communities. The villages visited in several field trips are in need of environmental remediation, protection of farm land from resource exploitation and decontamination of the environmental capital. Pollution, the lack of green infrastructure as well as clean sources of energy undermine the ecological, community resilience and the health of local people. Further explanation of the application of the transect in the remote mountain communities of this book are given in Section 3.0.; and examples of transect scenario constructions are shown in the Sections 3.3. – 3.5.

2.4. Fieldwork in the Mei Country Territory:

Key Findings from interviews and surveys with local people in the study area.

(To protect the privacy of all the interview participants their names have been redacted.)

This summary is based on interviews and surveys with over 35 people met in the study area. The interviews with locals where held in the study sites in Chiguang - Fengshuba Reservoir, Longbei – Heshui Reservoir and Changshazhen -Meijiang River. The purpose of the interactions with inhabitants in the communities was to have an understanding of the ways in which people would diversify their livelihood activities. The interviews with people revealed a limited awareness of environmental issues and sustainability. Conversations with people also gave insights of their knowledge of the ethnic background and local cuisine as a cultural asset for attracting visitors. The majority of people who participated in the surveys were local Hakka people. Some of the interviewee's parents were from the region, and the respondents were visiting family members, returning from work in larger cities including Guangzhou, Shenzhen and Hong Kong. Face-to-face interviews have been conducted with the help of a local contact Mrs. L. The interviews and surveys followed eleven topics, which included questions on Information sharing and communication, connectivity and transportation, food provision and preferences, water, energy sources, sources of income, e-commerce, knowledge on tourism and cultural resources, environmental and social consciousness. The surveys comprised of fifty-seven questions, including both closed-ended and open-ended questions. In total three-thousand-one-hundred and sixty minutes of interviewing time, has been spend while several of the surveys with local people where conducted over sharing dinner meals, tasting local cuisine, smoking cigarettes and sampling tea at Chinese Gongfu Tea Tables. Most of the respondents were knowledgeable in some of the question topics, and found it difficult to answer other questions of for them unfamiliar topics.

Practically, all interviewees were able to recommend local Hakka cuisine and specialties.

Information sharing:

In the study area mobile phone network coverage and internet penetration into the rural areas is available. While visiting smaller local shops, people were using smart phone payment methods such as Alipay and Wechat platforms. Information sharing with friends and family members has been noted as the next important way to obtain information updates, after the internet.

Connectivity:

Transportation by Car and trains is the prevalent mode of transportation. The majority of locals have a driver's license. Except for few people interviewed, most of the locals have taken airplanes for travels before.

Fig. 2.27: A local interviewee met in Longbei Town during the fieldwork phase of the research. The interviewee was collecting goods at the Taobao distribution centre in her town. Source: by the author.

Food provision:

Over half of the respondents expressed interest in growing their own food for personal consumption and for selling. Most of the respondents were not concerned of farmland loss due to the expansion of cities and construction activity. Several locals were aware of agricultural tourism and consolidations of small family farms by agricultural management companies.

Water:

All of the locals had sufficient drinking water supply for their households. Filtering of water purification has been noted by very few people. The digging of wells for sourcing of water is being practiced by some of the respondents.

Energy:

The prevalent sources of energy are gas and electricity power. Review energy sources used by few local people are solar water heaters installed on roofs and solar panel electricity generation. Approximately half of the interviewees expressed interest in installing clean energy sources for their own homes.

Income:

A typical average monthly income, excluding students, ranges between 3000-5000 Yuan (approx. USD 436 - 727.-).Several of the interviewees were self-employed. Respondents who were working in larger cities such as Guangdong and Shenzhen reported an average monthly salary of approx. 10000 Yuan (approx. USD 1454.-). Half for the locals responded they would prefer to stay with their community rather and seeking employment opportunities in different cities. Millennials (people born between 1981 and 1996) in this survey suggested making investments as an activity for additional sources of income.

E-commerce:

All of the locals interviewed were internet and e-commerce savvy. They use e-commerce platforms to sell and purchase goods. Several people noted the purchase of household goods, clothing, shoes and food on on-line platforms.

Most of the people had products delivered by small vans, none of the e-commerce users had any products delivered by drone transportation.

Local Tourism:

People recommended natural scenic areas, tea plantations, the Hakka Museum Park and Shenguang Mountain as their tourism assets. The introduction of local specialties such as Hakka dishes, Pomelo fruit and tea to visitors was recommended by many people. The interviewees noted the need of advertising and promotion of tourism into the region.

Environmental awareness:

Recycling of waste is largely not practiced. Inhabitants are not certain what measures local administrations undertake to reduce further water and air pollution. However, water pollution as an issue is in the consciousness of most of the respondents.

Social Services:

Access to primary school education is available to the communities. There is limited training provided for grownups, computer literacy courses and provision of re-training job skills in community centers. Elderly care is provided in nursing homes or care is taken by family members. Access to medicine is not a major concern; most of the locals obtain medication in drugstores or hospitals.

Preservation of Ethnic Culture:

Hospitality, frugality and hard work have been identified as some of the Hakka cultural characteristics. The continuity of Hakka culture through the teaching of mountain songs has been noted as part of the tradition. Hakka walled village, Yong Tau Foo (tofu stuffed with ground meat mixture and mushrooms) and Hakka lanterns have been mentioned as specific to the ethnicity.

Fig. 2.28: A liquor shop owner met in Chiguang Town during the interviews with inhabitants. Source: by the author.

Part Three

Research by Design Studies

The project as a knowledge producer.

3.0. Notes on 'Transect as a Method' as applied in the following design studies:

This introductory part summarizes how the transect as a method has been applied for the case study sites and design proposals in this research project. Equally, some of the limitations of the approach are mentioned in this section. For the purpose of providing clarification, the transect is not, exclusively, the only method used in this book to study or communicate ideas for regional design strategies. Two-dimensional ground plans drawings have also been used, predominantly, during the early stages of the research.

Larger scale ground plan drawings were produced, in particular, when I as the researcher was in the process of trying to manage the scale and complexity on an ambitious undertaking, while trying to record ideas and speculative solutions to the sustainable development issues found. Additionally, coming from a design discipline and architecture background, the transect as used in this research work has evolved through an iterative process. Iterative problem solving by design, as practiced by architects and planning professionals, or in other words through several processes of 'trial and error' and post-rationalizations, are understandably contestable for the social and natural sciences.

However, incrementally, working through the case study sites both in a Top-down and Bottom-up design approach, made me realize that the idealized 'smooth continuity' of ecologies and habitats as illustrated in the Geddes Valley section is rarely found in distressed territories. Instead, eco-system functions are disrupted or fragmented, due to human intervention and urbanization. Furthermore, ITC based New Geographies and transactional spaces give opportunities for small scale local interventions, which can have a large-scale impact transcending geographic boundaries.

Therefore, the transect tool during the later stages of the research evolved to combine the traditional advantages of working in sections through topographies, and the adding of localized interventions with new technological innovations, which did not exist for Geddes during his time.

The Transect as method:

In the context of the case studies undertaken in Part 3 of this book, which explore the transect as a method, planning interventions considering environmental remediation, village upgrading and measures aimed at improving the economic productivity have been proposed mainly for smaller sample sites in the case study villages.

The transect as a planning tool has been applied to understand: The topographic features of the sites, and spatial configuration of villages, the environmental issues including water and soil pollution, deterioration of the natural environment and the scarcity of arable land due to building construction or illegal resource extractions. The sectional method also enabled to visualize the relationships of the planning problems relative to each other in their spatial context. Additionally, acknowledging the importance of local social networks in the villages, direct interactions with community members helped to further identify the social issues, poverty and ecological degradation in the study sites. This lead to the inclusion of also micro-scale interventions in the transects considering the needs of the communities. Some of the inclusions are social welfare facilities such as schools, hospitals and community centers.

For the purpose of developing, regeneration and planning strategies for the fragmented landscapes, impoverished mountain communities and for establishing the interrelations with metropolitan regions, the transect as a tool can be scaled to incrementally conduct planning work in micro and macro - scale regions. Although an ambitious regional planning challenge, the approach of embedding remedial measures as studied in the transects (Fig. 3.97a –c) could be applied to reinstate continuities of natural resources networks such as river networks and larger scale ecological corridors. Further, research into the possibilities of the transect as research and design tool would include, utilizing the transect as a framework for structuring research of remote mountain to mega-city regions to manage the complexity of regional relationships at such scales.

Limitations of the approach:

Certainly, for both sectional and plan based studies, the limitations of incorporating an architect's idiosyncratic perspective and a 'trial and error' problem solving approach by designing, are prone to subjective decision making, common for the design disciplines. Further, the difficulty of testing the design proposals on site leaves a planner to his or her judgment, unless the planning work is informed by a multi-disciplinary team of specialist consultants.

In particular, for regional scale problems, where proposed solutions take several to realize and to re-evaluate after the realization of planning schemes, the effectiveness of development or regeneration projects can be subject to political or economic cycles, for example.

The difficulty of proposing solutions to regional scale issues by also looking at micro-scale interventions in the communities and for individuals, as in this study, undoubtedly, makes the design problems difficult to solve. Further it is difficult to relate micro-scale interventions to mega-city regions. Nevertheless, trying to give alternatives for a people - oriented new type urbanization by working only at the scale of mega-city corridors, gives only a high-level, macro-scale strategic approach to addressing regional planning problems.

Nonetheless, an advantage of the transect is that it allows planners to understand hydrological and geological conditions in sections, in relationship to large-scale topographical variations. Further, the transect as a tool and a geographic unit is scalable. Therefore, despite challenges of complex transactional networks at 'remote mountain to mega-city scales', the transects as a tool with 'additions' remains relevant for regional and community level research.

Research by Design Studies - The project as a knowledge producer.

Part Three is the research by design part of this project. The scenario constructions in the last Parts 3.5 (see sections shown in Fig. 3.97a – 3.97c) and 3.6 synthesize the analytical mapping work and propositional design explorations presented in following subsections. The scenario constructions proposed in Section 3.4, Characters 01 -05 (Livelihood diversification strategies of local inhabitants) are then related to the scenario constructions in Part 3.5 as exemplified in the above mentioned axonometric sections. The final scenario constructions suggest ways of embedding new grass-roots level livelihood strategies to empower individuals and the support needed to set up socially-conscious welfare facilities in the communities. Ultimately, Part 3.6 presents examples for Street Transformations in the three study settlements Chiguang town, Longbei town, and Changsha town in Mei County.

This research by design section has been structured in Four Scales of investigation. Scale 01 provides a macro-scale regional analysis and contextualization. The Greater Bay strategic vision with the Pearl River Delta merging into a mega-city cluster is shown. The marginal settlements mapped out in Section 3.3 Scale of Investigation 03, located at water supplying reservoirs for the mega-city, are not considered by the government's official regional master plans, which overemphasize the development of metropolitan areas.

As the Scale 02, the maps collect analytical information about the Dongjiang river basin territory in Section 3.2. Based on the analysis on the territory, Sections 3.2.1 and 3.2.2 explore and propose alternative scenarios in two-dimensional regional plans and sections. The scenarios tackle the following issues: Proposing a Green Infrastructure to purify water and soil contamination of the hydrological system, alternative scenarios for the preservation of basic farm land for food self-sufficiency, re-settling the people in the upstream region of the territory in low-density settlements away from the overpopulated delta region, the idea of developing new cities in the upper zones of the territory to catalyze economic growth, utilizing steeper topographies in the mountains to cultivate forests, and the capitalizing of the natural scenic areas in the mountains and the Hakka ethnic traditions for the development of the local tourism industry. The proposals in this section are 'Top-down' design approaches; whereas the following scales are "Bottom-up" design approaches or grassroots level interventions.

The Scale 03 explorations look at three marginal settlements along a section line in Mei County. The section line is cut from the mountainous Fengshuba reservoir to the urban fringe area of the Capital of Hakka Culture Meizhou. As the Hakka cultural capital is instrumental to the tourism potential of the marginal settlements, it is signification to maintain this ethnic relationship and affiliation for the study areas. The three study areas are Chiguang town - Fengshuba Reservoir, the most peripheral and therefore, poverty-stricken settlement in the Upstream region, Longbei town – Heshui Reservoir a 'Taobao Village' example in Mei County, and Changsha town along Meijiang River, a peri-urban settlement with mixed agricultural and industrial activities. Additionally, the Yannafei tea plantations, of this region are mentioned, as it is a local agro-tourism example which due to its popularity by visitors cannot be overlooked.

Finally, at Scale 04 are the five micro-scale scenario constructions (see Fig. 3.89 – 3.93) proposed based on fieldwork research and interactions with local inhabitants. To overcome poverty the five characters diversify their fields of livelihood activity into Hakka cuisine & organic farming, Eco-lodge hosting, Smart Agronomy & Sustainability education, Tea guesthouse hosting & community centre, new transportation mobility, Temple venue guide, Chinese medicine produce , part-time farming, bus & courier driver, and Agro-Tourism – tour guide. The grass-roots level livelihood scenarios have been proposed considering the aspirations of the New Type Urbanization Plan in China.

3.1. Scale of Investigation – 01.

Satellite view of the South-East Coastal zone.

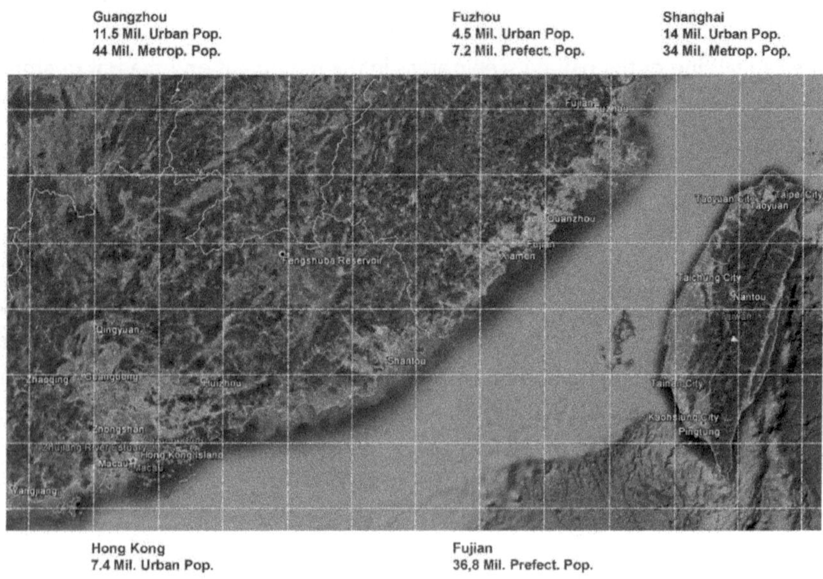

Fig. 3.00: The satellite views shows part of Guangdong province, Jiangxi and Fujian provinces. Source: Google Earth, June 2018.

Regional Context: As part of the Belt & Road initiative, the Guangdong - Hong Kong - Macau Greater Bay Area was signed between the National Development and Reform Commission (NDRC) and the governments of Guangdong, Hong Kong and Macau in July 2017 with Chairman Xi Jinping's presence. This ambitious plan aims to transform the Pearl River Delta (PRD) including Hong Kong and Macau to a dynamic hub of innovation and services with a GDP of US$4.62 trillion by 2030.

Fig. 3.01: New Synergies of Hong Kong's financial and professional services sectors, Shenzhen's high-tech manufacturing and innovation skills, and the manufacturing strengths of Dongguan and Guangzhou. The development of the area should also act as a catalyst for China's Belt and Road initiative–an ambitious strategy that aims to link the economies along the Silk Road Economic Belt (Central Asia to Europe) and the Maritime Silk Road (South Asia to Africa and the Middle East) together. Source: Fung Business Intelligence, 2016.

Policy Collaborations: The Guangdong – Hong Kong – Macao Greater Bay Area comprises the two Special Administrative Regions of Hong Kong and Macao, and the nine municipalities of Guangzhou, Shenzhen, Zhuhai, Foshan, Huizhou, Dongguan, Zhongshan, Jiangmen and Zhaoqing in Guangdong Province.

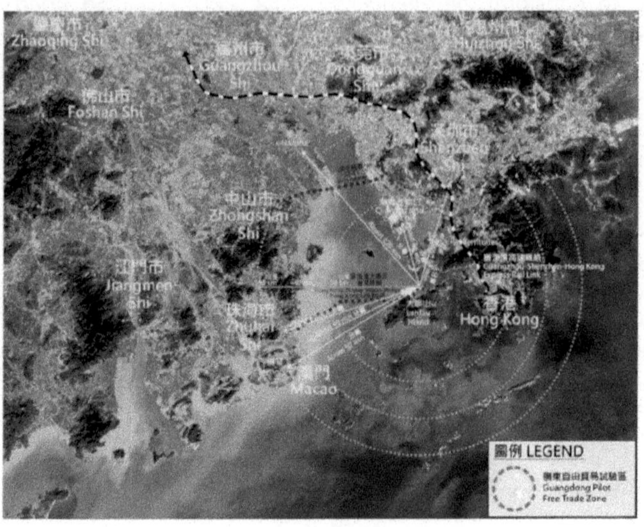

Fig. 3.02: The coordinated economic development of the Greater Bay Area is to maximize the competitive strength of all cities in the region. The Pearl River Delta is seen as key region for China's innovation driven development. The areas of collaboration include (1) Innovation and Technology, (2) Financial Services, (3) Transportation and Logistics, (4) Professional Services, (5) International Legal and Dispute Resolution Services, (6) Clearance Facilitation, (7) Medical Services, (8) Education, (9) Art, Culture and Creative Industries, (10) Tourism, (11) Environmental Protection and Sustainable Development, (12) Youth Development. Source: Develop Bureau, Planning Dept. Oct 2016;www.bayarea.gov.hk.

"The idea is to create a 'one hour living zone' encompassing all the nine cities in the Pearl River Delta,"

"Such interconnectedness among the nine cities is beyond the conventional urban agglomeration,"

... *"Instead of considering them to be nine individual, separate cities, it makes more sense to see them as a unified, organic whole with a continuous urban area. It's fitting to call it a megalopolis."* Prof. Zuo Zheng, in Economics, Jinan University.

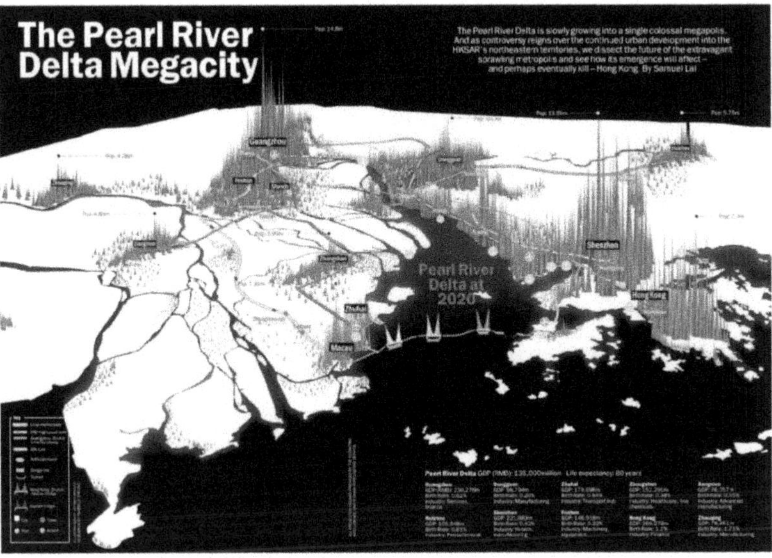

Fig. 3.03: The Pearl River Delta is gradually merging into a single mega-city. Source: Samuel Lai, TimeOut HK, 2016.

Top-down Cross-Boundary Development Cooperation of the Greater Pearl River Delta.

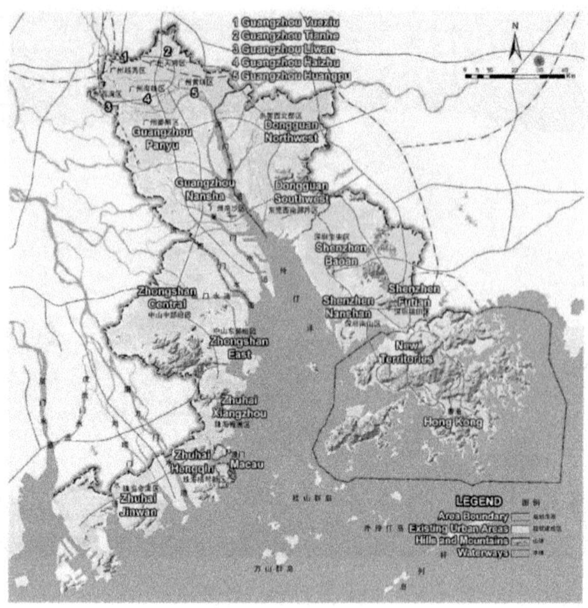

Fig. 3.04: Overall Territory: Action Plan for the Bay Area of Pearl River Estuary. Source: Hong Kong Planning Department, 2011.

Creation of Central Business Districts as Infrastructure for Participating in an Internationalized Economy and Global City Network. The Dongjiang River Basin is not included in this system of command centers.

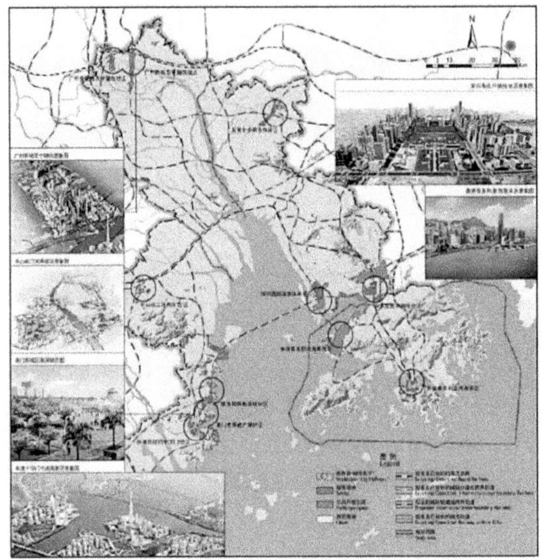

湾区"城市客厅"重点地区示意图
World-class "City Hallways"

Fig. 3.05: Key City Hallways: Action Plan for the Bay Area of Pearl River Estuary. Source: Hong Kong Planning Department, 2011.

The Transport Network is mainly centered around the connectively in the Pearl River Delta Estuary. Linkages into the hinterland which is supplying resources to the megacity is not shown in the plan.

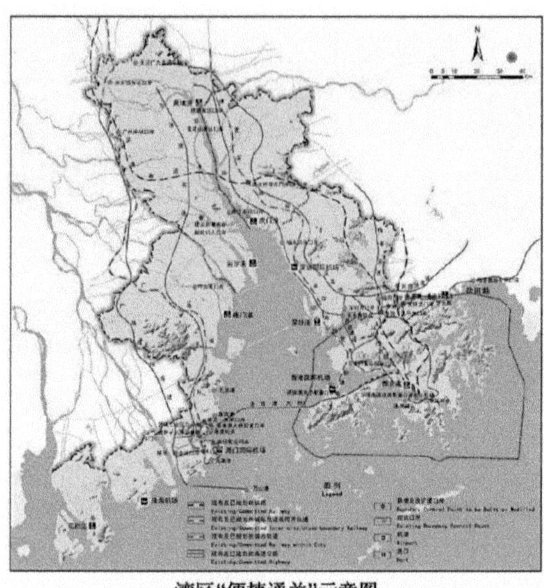

湾区"便捷通关"示意图
Cross-boundary Transport System

Fig. 3.06: Cross-boundary Transport system: Action Plan for the Bay Area of Pearl River Estuary. Source: Hong Kong Planning Department, 2011.

The Interdependency and Full extent of Ecological Infrastructure servicing important ecosystem functions including absorption of pollutants, water provision and farm land is beyond the scope illustrated in this map.

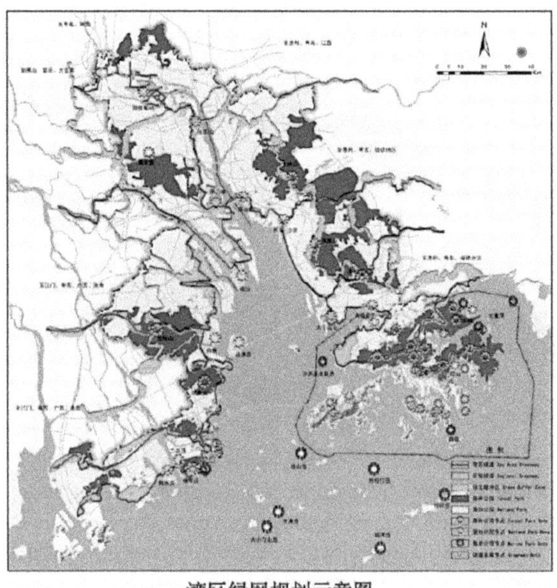

Fig. 3.07: Green Network: Action Plan for the Bay Area of Pearl River Estuary. Source: Hong Kong Planning Department, 2011.

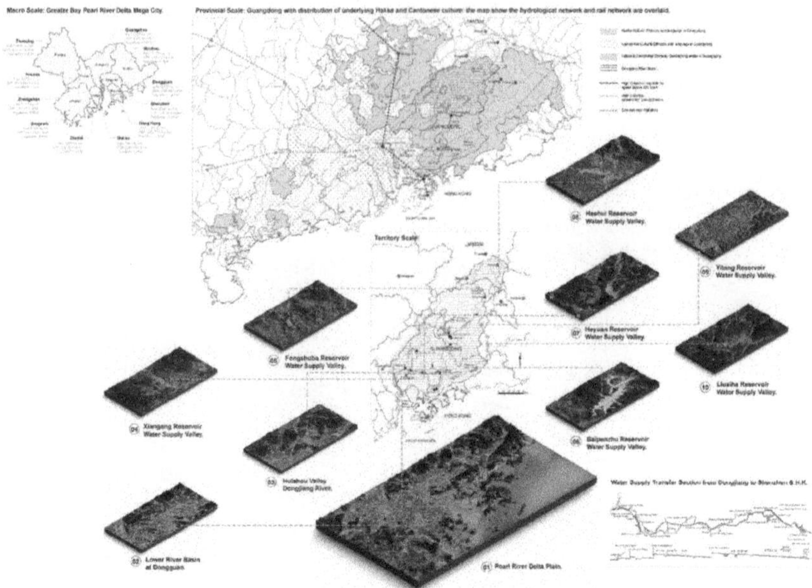

Fig. 3.08: This illustration shows the water supply reservoirs of the PRD mega city in a consolidated axonometric drawing. At a macro scale, the drawing intends to synthesize the underlying Hakka cultural ethnicity, the hydrological capital and rail network linkages found in the region. Due to the overlapping nature of regional ecological, cultural and infrastructural assets, illustration exemplifies the difficulty of determining definitive borders for the regional features, which transcend geographic boundaries found in ordinance survey maps. The small DEM (digital elevation model) of the reservoirs are enlarged in the following pages for better viewing. Sources: Google Earth & Image generated by the author, 2019.

Fig. 3.09: Digital Elevation Model 01 Pearl River Delta Plain. Refer to Fig. 3.08 for context. Sources: Google Earth & Image generated by the author, 2019.

Fig. 3.10: Digital Elevation Model 02 Lower River Basin at Dongguan. Refer to Fig. 3.08 for context. Sources: Google Earth & Image generated by the author, 2019.

Fig. 3.11: Digital Elevation Model 03 Huizhou Valley Dongjiang River. Refer to Fig. 3.08 for context. Sources: Google Earth & Image generated by the author, 2019.

Fig. 3.12: Digital Elevation Model 04 Xiangang Reservoir Water Supply Valley. Refer to Fig. 3.08 for context. Sources: Google Earth & Image generated by the author, 2019.

Fig. 3.13: Digital Elevation Model 05 Fengshuba Reservoir Water Supply Valley. Refer to Fig. 3.08 for context. Sources: Google Earth & Image generated by the author, 2019.

Fig. 3.14: Digital Elevation Model 06 Baipenzhu Reservoir Water Supply Valley. Refer to Fig. 3.08 for context. Sources: Google Earth & Image generated by the author, 2019.

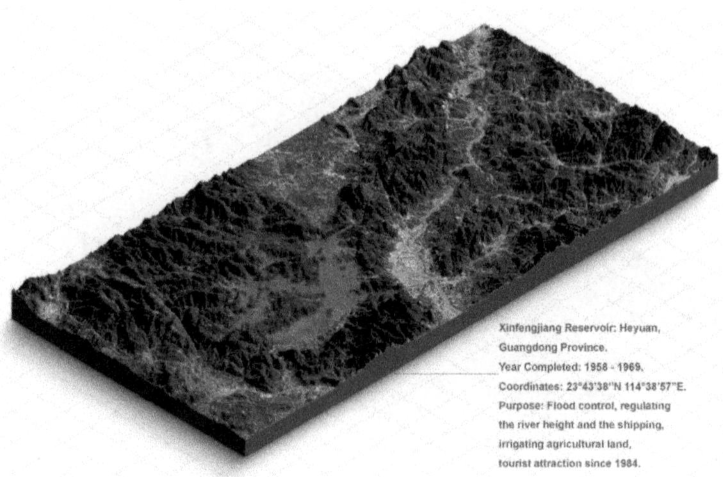

Fig. 3.15: Digital Elevation Model 07 Heyuan Reservoir Water Supply Valley. Refer to Fig. 3.08 for context. Sources: Google Earth & Image generated by the author, 2019.

Fig. 3.16: Digital Elevation Model 08 Heshui Reservoir Water Supply Valley. Refer to Fig. 3.08 for context. Sources: Google Earth & Image generated by the author, 2019.

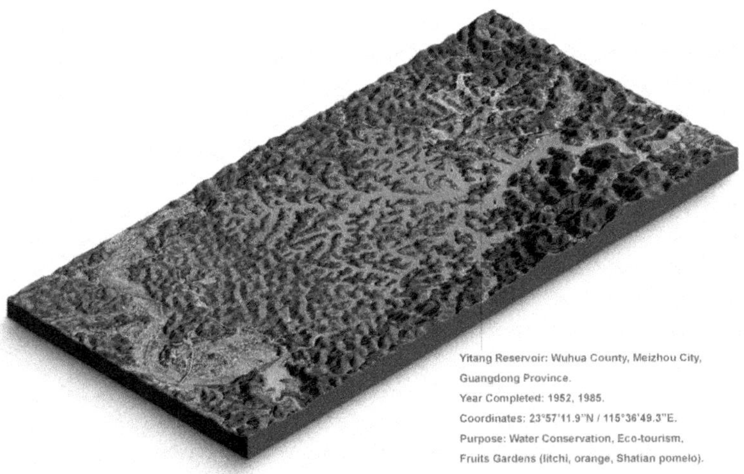

Fig. 3.17: Digital Elevation Model 09 Yitang Reservoir Water Supply Valley. Refer to Fig. 3.08 for context. Sources: Google Earth & Image generated by the author, 2019.

Fig. 3.18: Digital Elevation Model 10 Liuxihe Reservoir Water Supply Valley. Refer to Fig. 3.08 for context. Sources: Google Earth & Image generated by the author, 2019.

3.2. Scale of Investigation – 02.

Mapping & Analysis of the river basin territory.

Hydrological Network and Water resource – Analysis of the rivers and Reservoirs of the Dongjiang River Basin.

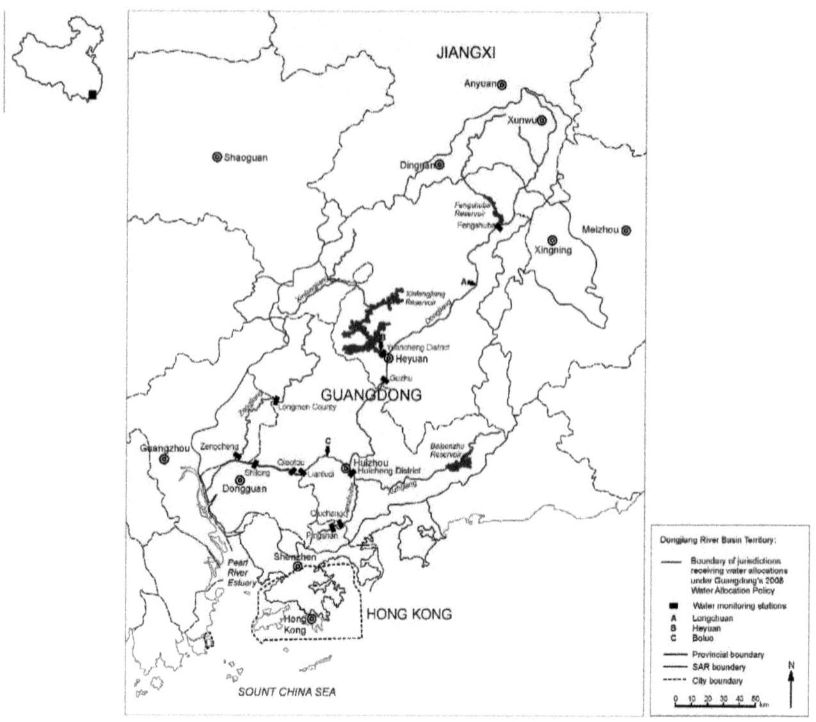

Fig. 3.19: Hydrological Network – Dongjiang River Basin. Source: by the author, based on map by Depart. of Water Resources of Guangdong Prov. (DWRGP).

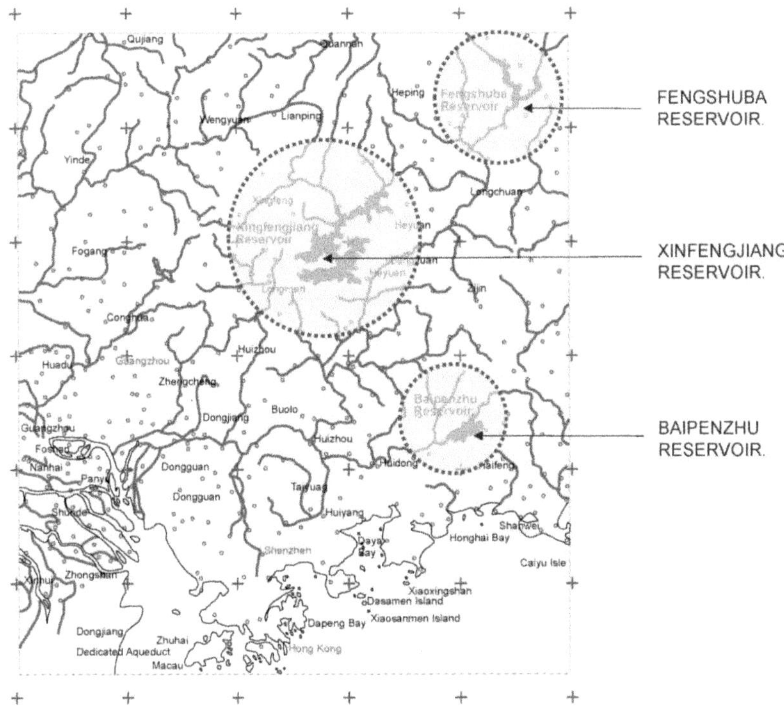

Fig. 3.20: Map of the three main reservoirs in the Dongjiang river basin. Source: by the author, Main Water Resources of the Dongjiang River Basin: Fengshuba, Xinfengjiang, and Baipenzhu Reservoirs; based on map by Depart. of Water Resources of Guangdong Prov. (DWRGP).

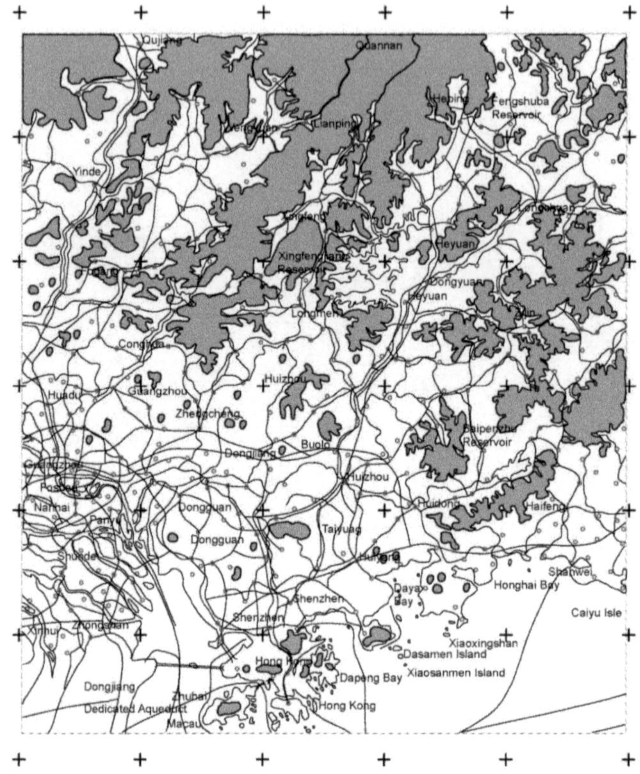

Fig. 3.21: Major Concentration of Green Open space and Forestry reserves – Dongjiang River Basin Territory. Green Open space and Forestry Enclaves – Dongjiang River Basin. Source: by the author, based on map by Depart. of Water Resources of Guangdong Prov. (DWRGP).

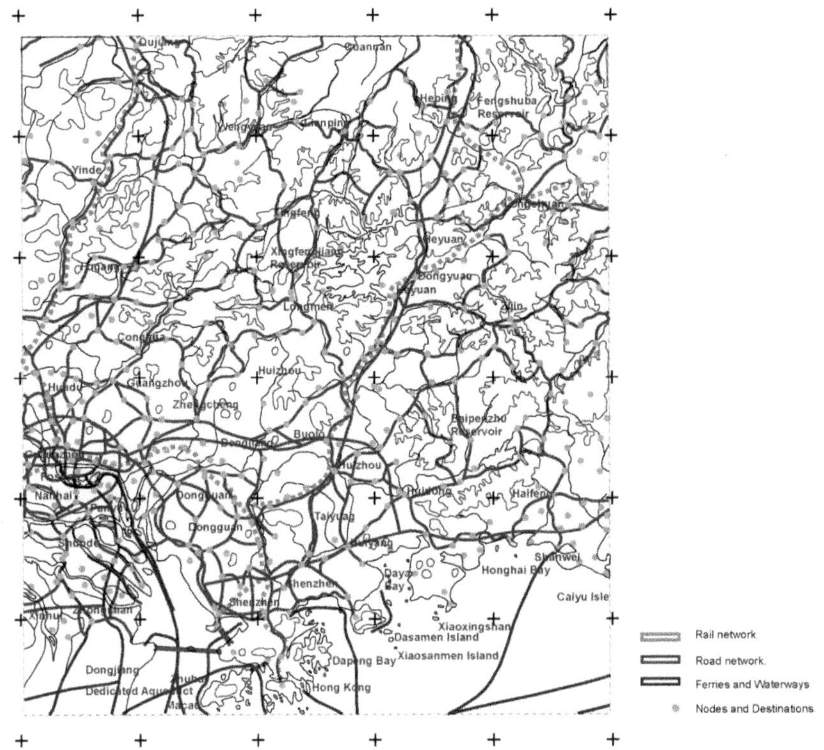

Fig. 3.22: Transport Infrastructural Network and destinations – Analysis of the existing System of Mobility. Transport Infrastructural Network – Dongjiang River Basin. Source: by the author, based on map by Depart. Of Water Resources of Guangdong Prov. (DWRGP).

Maps of Land use and Soil quality indices:

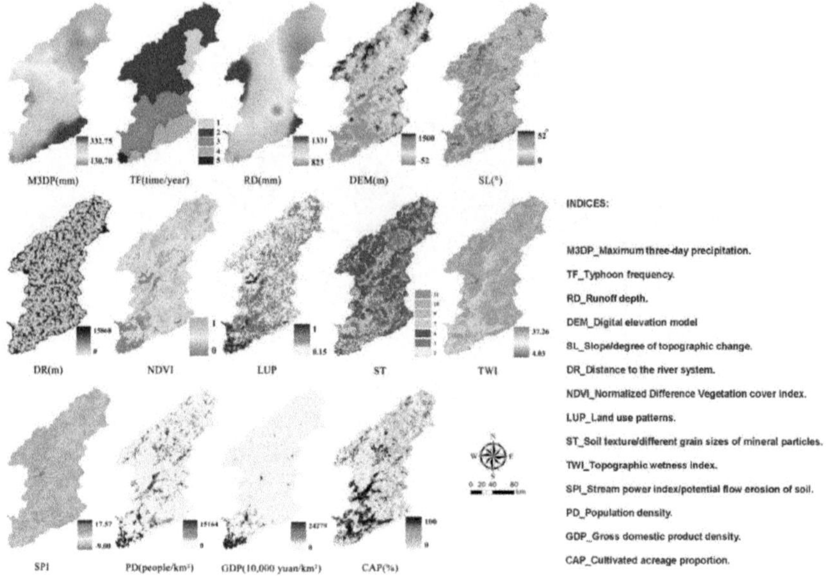

Fig. 3.23: Maps of Land use and Soil quality in the Dongjiang river basin. Source: Lai, Shao, Chen, Wanga, Zhou, Yang, Zhang, 2016.

Upstream, Midstream and Downstream zones of the River network:

The Dongjiang River Basin extends from the mountainous border region of Jiangsu and Guangdong Province into the Pearl River Delta Estuary.

The map shows the subdivisions of the river basin into nine sub-catchment areas.

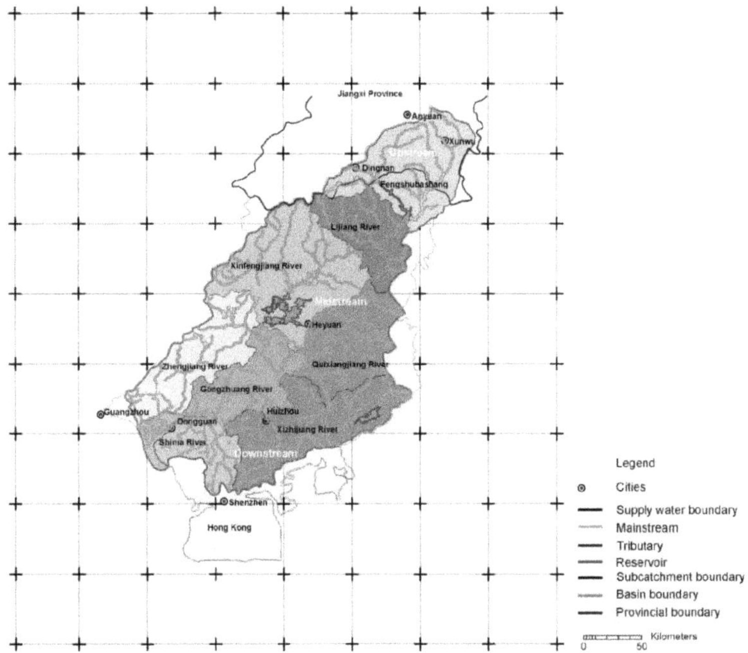

Fig. 3.24: Structure of the river basin. Source: by the author, based on Lee, 2014.

Flood Risk: classification into six risk levels.

Climate change, extreme weather conditions and sea level rise increasingly put coastal and river delta regions at the risk of flooding. Demarcations into six flood risk levels in the Dongjiang River Basin range from lowest risk to highest risk level. The downstream delta region is at the highest risk of flooding. The mountainous Mei Country region in the upstream zone of the river basin has the lowest risk level and therefore may have future advantages for new settlements.

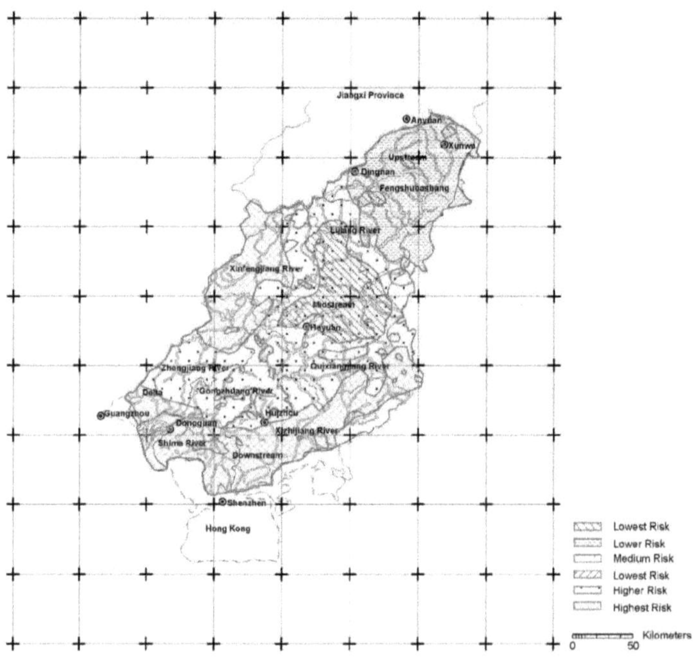

Fig. 3.25: Flood risk map. Source: by the author, based on Lai, Shao, Xiaohong, Wang, Zhou, Yang, Zhang, 2016.

Types of existing industries contributing to contamination of river network.

The contamination of the hydrological river network is shown in five Grades I – V. The Dongjiang main stream river has a medium level of contamination, or Grade IV. The main stream river water is suitable for industrial and agricultural use only. Future programs for industrial relocations by The People's Government of Guangdong Province are being discussed by the government to prevent further deterioration of the water quality in the basin.

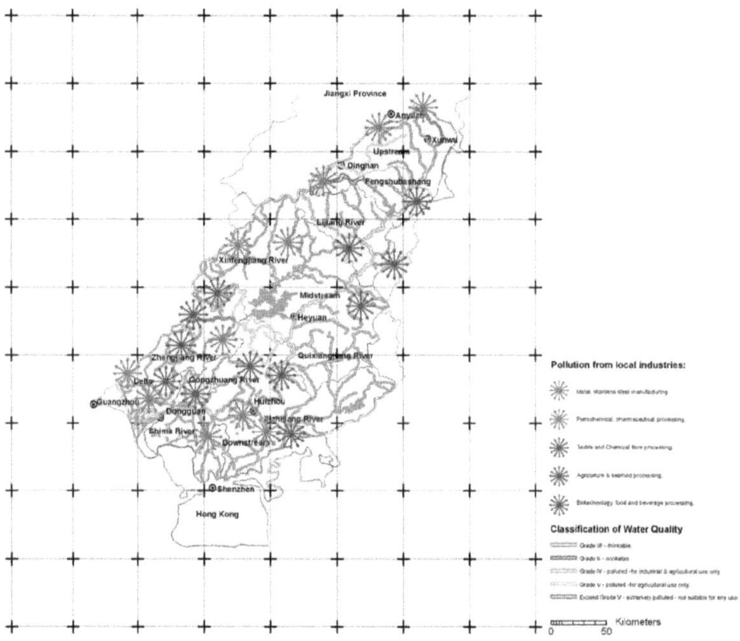

Fig. 3.26: Analysis of types of contamination. Source: by the author, based on Liu, 2012.

Topographic Elevation with 'sub-alpine' mountain territories:

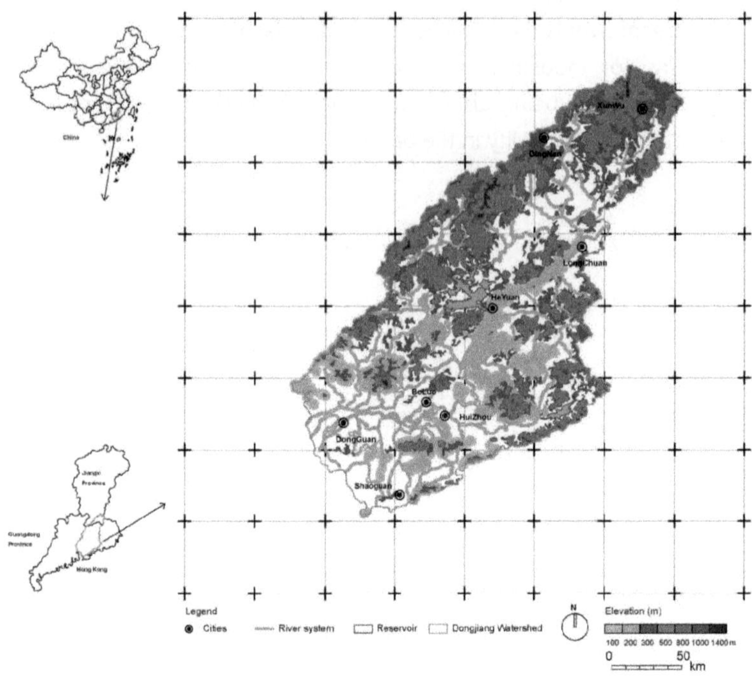

Fig. 3.27: Elevation map 100 - 1400m.Source: by the author, based on Ding, Jiang, Fu, Liu, Peng, Kang, 2015.

Currently 'urbanized areas' by official land use classification:

The total population of the river delta and urbanized regions is approx. 46 million people. The main cities considered for the river basin include Guangzhou, Huizhou, Dongguan, Shenzhen and Hong Kong. On average the population density in the downstream delta region is > 900 people / km², whereas the average population density in the upstream Mei County region is approx. 100-200 people / km².

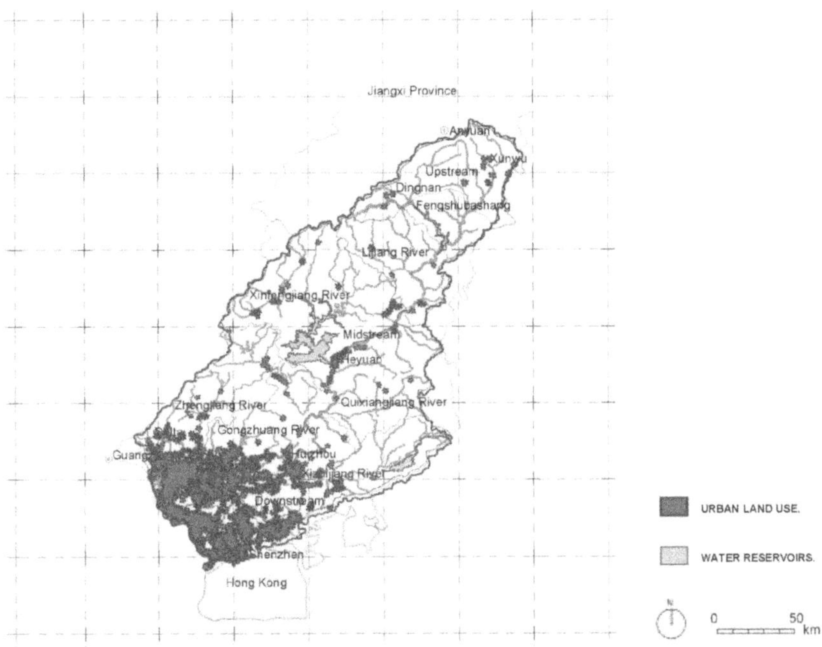

Fig. 3.28: Urban land use based on administrative boundaries. Map by the author, based on data by Ding, J., Jiang, Y., Fu, L., Liu, Q., Peng, Q., & Kang, M. (2015).Impacts of land use on surface water quality in a subtropical River Basin: a case study of the Dongjiang River Basin, Southeastern China. Water, 7(8), pp. 4427-4445.

Current Agricultural land use:

Currently, less than 9000 km² is dedicated to industrial-scale agricultural land use in the territory. While small family farms cultivate plots of land predominantly to sustain their personal food consumption, modern agricultural knowledge is necessary to increase the food production efficiency also for smaller agricultural operations.

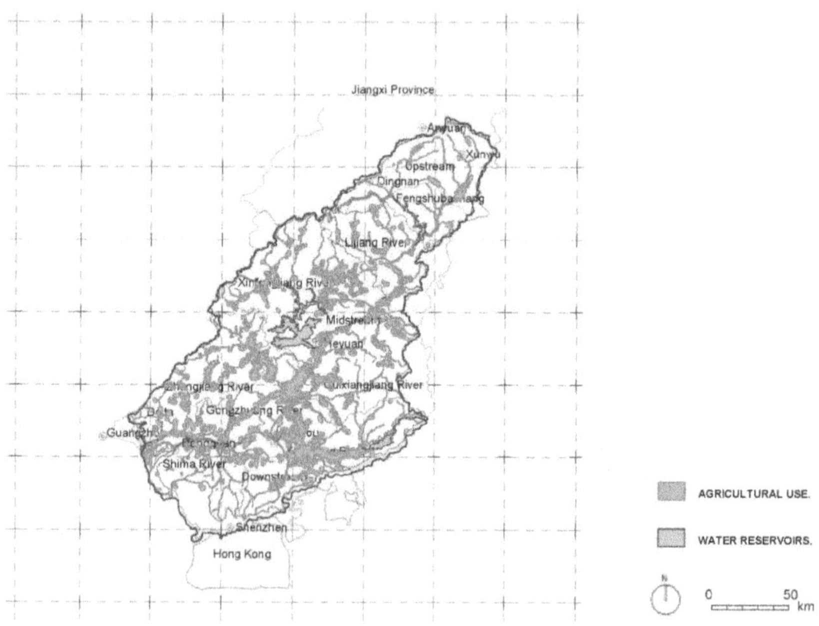

Fig. 3.29: Agricultural land use. Map by the author, based on data by Ding, J., Jiang, Y., Fu, L., Liu, Q., Peng, Q., & Kang, M. (2015).Impacts of land use on surface water quality in a subtropical River Basin: a case study of the Dongjiang River Basin, Southeastern China. Water, 7(8), pp. 4427-4445.

Map of Soil Fertility:

The loss of fertile land to new urban expansion is an issue in China compromising independent crop production. To reduce the reliance on food imports from foreign countries, China has passed the Basic Farmland Protection Regulation in 1994. The policy designates basic farmland protection districts for townships. Further, the regulation prohibits the conversion of basic agricultural land for housing, infrastructure or industrial use. Originally, the fertile soil in the delta and coastal regions has been lost to the development of export-focused cities in the coastal region.

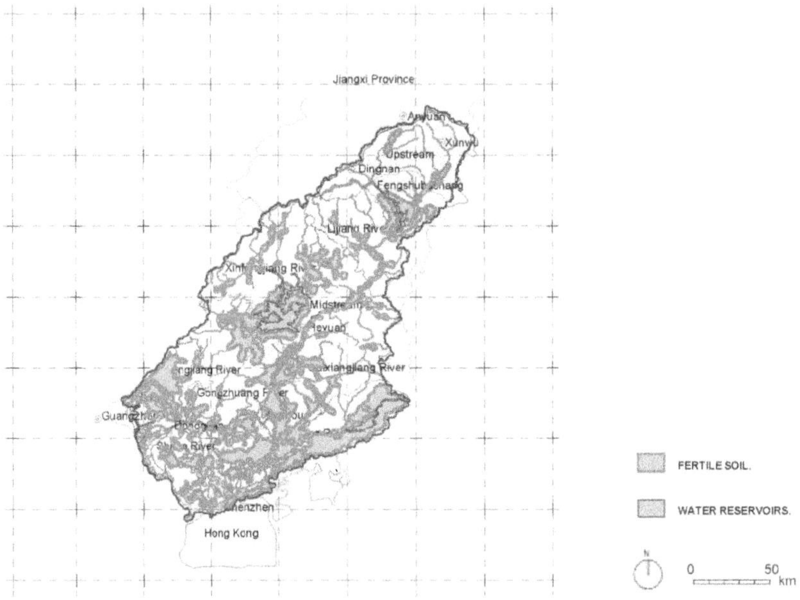

Fig. 3.30: Map of soil fertility. Map by the author, based on data by Ding, J., Jiang, Y., Fu, L., Liu, Q., Peng, Q., & Kang, M. (2015).Impacts of land use on surface water quality in a subtropical River Basin: a case study of the Dongjiang River Basin, Southeastern China. Water, 7(8), pp. 4427-4445.

Map of GDP per capita income level and the concentration of acute poverty.

The degree of poverty magnifies as the geographic distance from the economically developed Pearl River Delta increases. The dark brown regions shown in the map are the poverty-stricken mountain areas in Mei County. Larger-scale strategic interventions are proposed in the following studies. At the scale of communities and individual people, the possibilities of new diversified livelihood strategies for communities in mountain regions are explored as scenario studies in the subsequent Parts 3.4, 3.5 and 3.6.

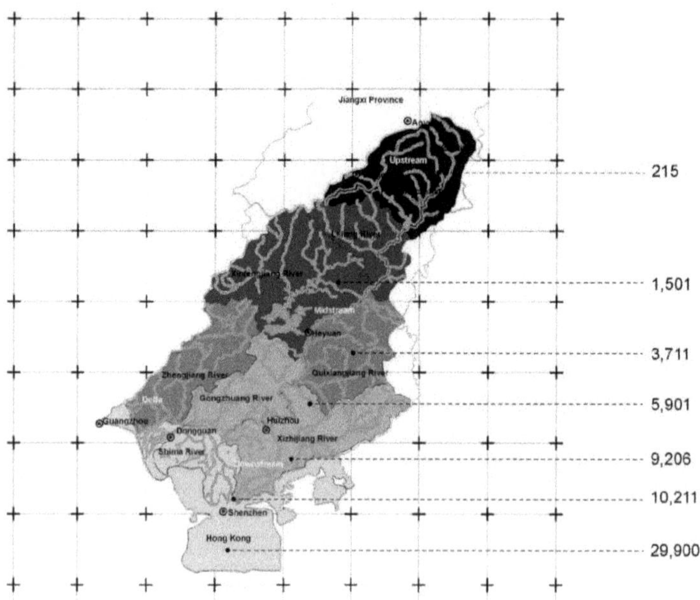

Fig. 3.31: Per capita income level in USD/p.a. Source: Carlos Lo, 2010.

3.2.1. Dongjiang River Basin Scenarios – a Top down approach.

Scenario studies and propositions.

Ecological Remediation zone for polluted Hydrological network and renewable technologies.

At the scale of the Dongjiang River basin, this scenario prioritizes the decontamination of the Dongjiang River. The intervention is proposing an ecology remediation strip along the primary river streams. A green strip with phytoremediation vegetation is planted at the river banks to extract pollutants from soil and water. Some of the plant species may include reed beds, tall fescue, basket willow and sunflower fields.

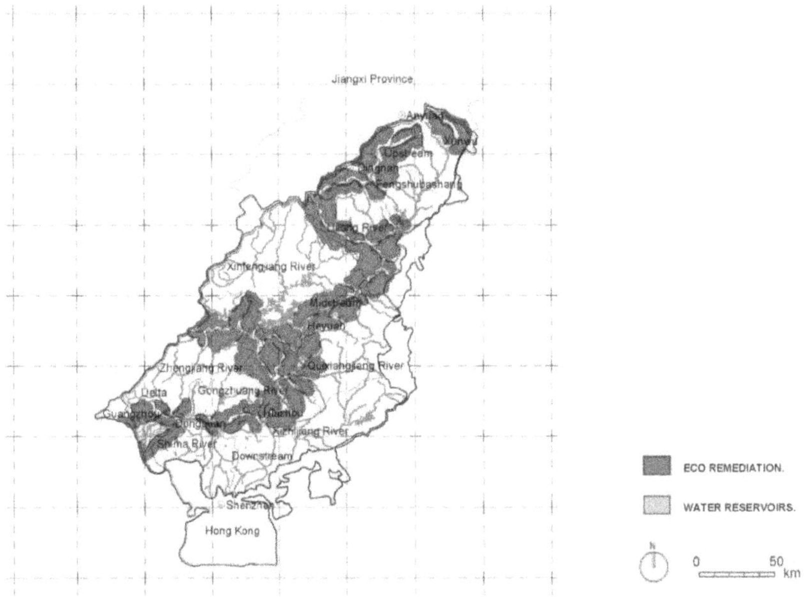

Fig. 3.32: Scenario for an ecology remediation strip along highly polluted river network. Image Source: by the author.

Hydrological network pollution: Dongjiang River Basin with Eco-remediation Zone superimposed.

The map shows the hydrological network contamination map (refer to Fig. 3.26) from the river basin analysis part with an overlay of the eco-remediation green strip. The newly created wetlands as ecology parks are intended to prevent the development of land for urban construction. The wetland riverbanks can overtime evolve into wildlife habitats and recreational scenic areas.

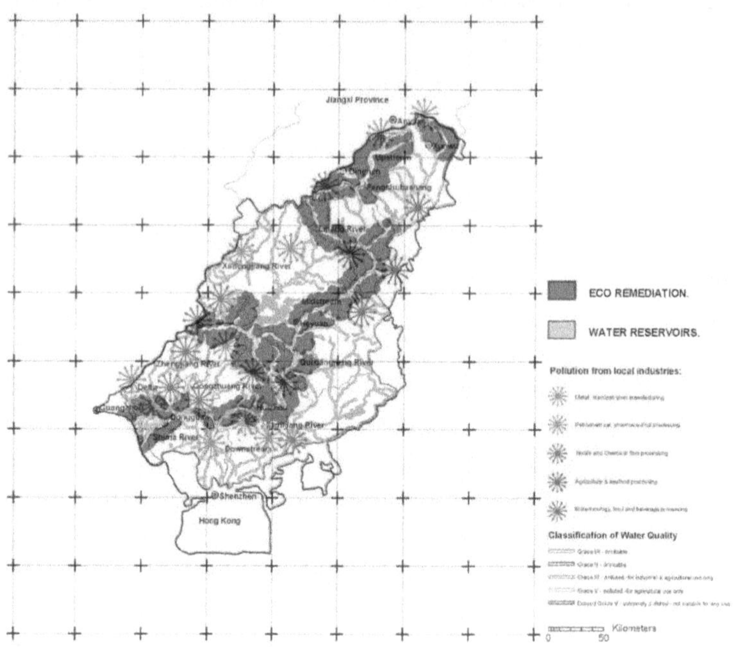

Fig. 3.33: Eco-remediation zone shown with classification of water quality. Image Source: by the author.

Scenario for an increase of existing Agricultural collectives by 30%. Total farm land area growth to 11250km2.

The current agricultural land use as shown in basin analysis part (see Fig. 3.29) has been taken as a starting point. This scenario envisions a gradual increase of agricultural land by expanding farm land at the perimeters of existing agricultural land. The idea is to incrementally re-appropriate adjoining plots of land in proximity to farming operations in the territory. This strategy aims at securing land for food production.

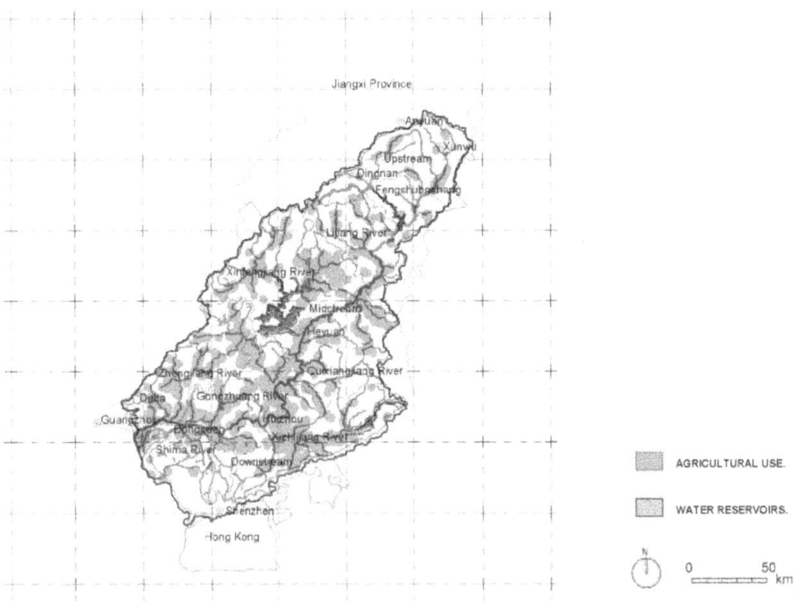

Fig. 3.34: Food security: more agricultural production by increasing the farm land of the existing agricultural operations. Image Source: by the author.

Basic farmland protection zone for food independence as a territory In-between. Approximate agricultural land allocation of 11250km2. The scenario proposes to minimize the further loss of farm land and food insecurity due to rapid urbanization.

Urban development corridors, road and high-speed rail networks create divisions in the river basin. At the regional scale, the spaces in-between the urbanization corridors can be preserved as land for large-scale consolidated farming industries. The agricultural consolidation as shown in this scenario limits the further expansion of the Pearl River Delta Mega city into the upstream regions of the territory.

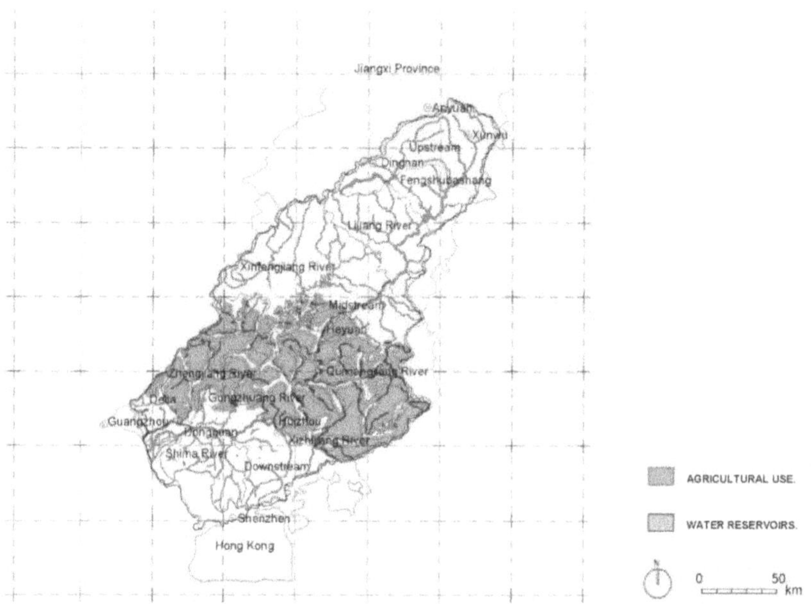

Fig. 3.35: The map suggests a farm land protection zone in-between the mega-city and the mountainous hinterland to maintain independent food production capacities. Image Source: by the author.

Dispersed urbanization scenario: 'Isotropic organization', or even distribution of low density settlement structures outside the main urban agglomerations.

This map shows the concentration of urban land use as the red filled-hatch, also see in the river basin analysis part (Fig. 3.28). The settlement of people away from the overpopulated urban agglomeration is a necessity for addressing the 'floating' population issue in Chinese cities. This scenario suggests the setting up of low-density settlements in the mid and upstream regions of the basin as indicated as a finer red hatch. This is to avoid the negative impacts to the environment caused by uncoordinated development and overpopulation.

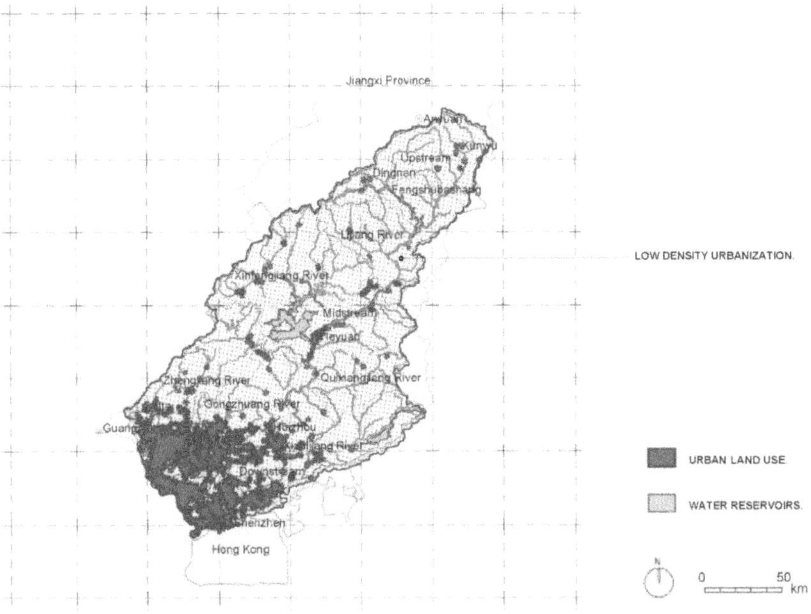

Fig. 3.36: The finer grid in red shows distribution of low-density settlement structures. For the sources of the concept of Isotropy in urbanism refer to: Viganò, P. (2008). Water and Asphalt The Project of Isotropy in the Metropolitan Region of Venice. Architectural Design, 78(1), 34-39. Image Source: by the author.

Mountain Territories as a resource rediscovered. Scenario of a new type, low density urbanization in the 'sub-alpine' hinterland.

Two of the scenarios considering proposals for the resettlement of people in mountainous regions and the protection of basic farm land for food security are shown in combination. Refer to the interventions presented in isolation in Figures 3.35 and 3.36. This scenario suggests a reduction of the population in the mega city region to ameliorate the burden to the metropolitan environment. The scenario proposed sustainable low-density settlements outside the downstream urban agglomeration.

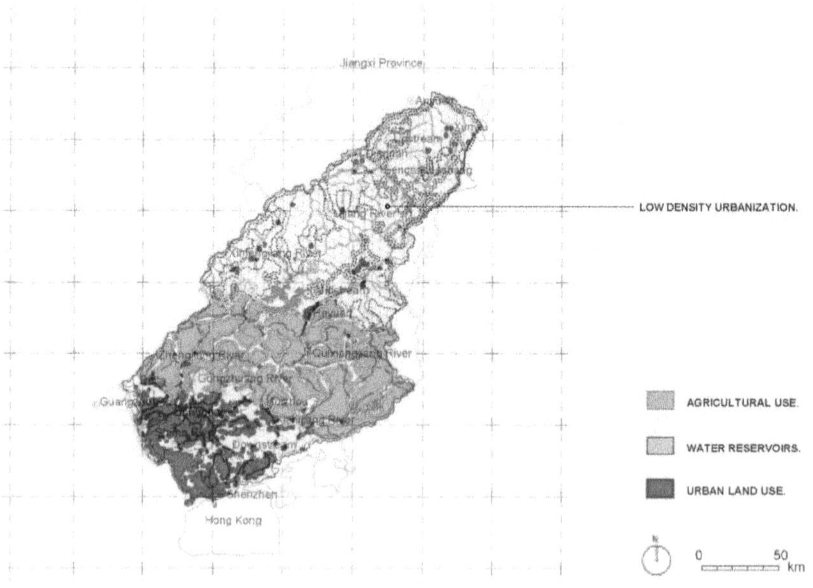

Fig. 3.37: This scenario maintains the megacity region, while securing a consolidated farm land protection zone and dispersed urbanization in the mountains. Image Source: by the author.

Proliferation of mixed 'rurban' informal industrial and agricultural activities, as indicated in fine red hatch.

In this scenario, the proposal maximizes the cultivation of fertile soil as mapped out in the basin analysis part (refer to Fig. 3.30) for the purpose of increased agricultural productivity. This radical proposal utilizes available fertile soil also in the downstream urban agglomeration. The scenario implies a reintroduction of agricultural activities in urban areas, and encourages farming activities by citizen in the metropolis. By reclaiming fertile soil in urban areas for agriculture, the scenario suggests the redistribution of the urban population to the mid and upstream regions of the basin into new low-density settlements.

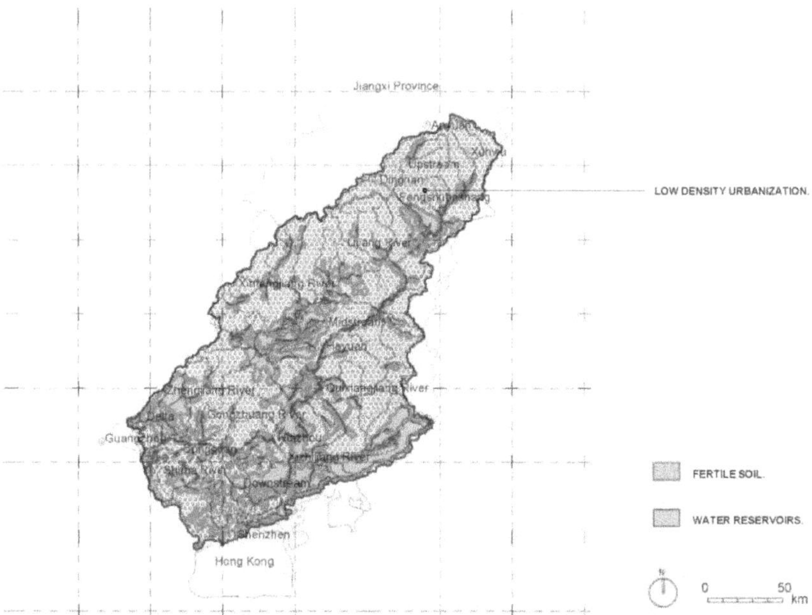

Fig. 3.38: While maintaining the existing agricultural ecologies, this scenario proposed a horizontal distribution of economic ecologies which include both rural and urban livelihood activities. Image: by the author.

New Towns of One Million Inhabitants each, to counteract uneven urbanization and inequality in the river basin.

For the purpose of clarification, please note that this study, as a 'Top-down' design approach, obviously, represents a different perspective and provides contrast to 'Bottom up' or 'grassroots level' design interventions in communities. While small-scale studies and possible improvements for local inhabitants are presented in Fig. 3.97a-3.97c. , for the planning in larger scale territories, neither 'Top down' nor 'Bottom up' design approaches can be excluded. Therefore, the need for designing at a series of scales is considered.

The development of new cities of approximately one million people in the mountain regions is suggested to reduce the overpopulation burden in the delta region. This proposal combines the idea of a green strip along the Dongjiang river and the consolidation of agricultural land use in the mid-stream region between the downstream mega city and new cities in the mountain areas.

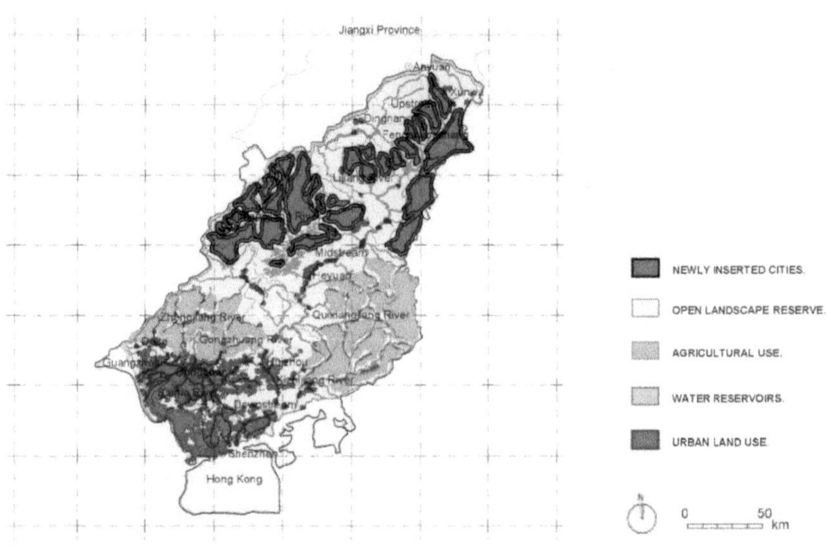

Fig. 3.39: The scenario proposes the insertion of small cities in the economically disadvantaged territories. It suggests the resettlement of a displaced migrant population from larger cities into new communities. The setting up of new communities would require support from the government and the need to attract investment capital. Large-scale agricultural companies in partnership with Provincial and Township level government could collaborate in development associations to attract the necessary funds for urbanization.

Scenario of a Mega-city decline and 'sub-alpine' territories as new alternative cities.

In an event of a collapse of the mega city, the new cities designed, based on people-oriented &environmentally friendly development principles offer an equitable alternative to living and working in the mega city region. Due to the relatively smaller population of one million, as a habitat, alternative cities in mountain regions may become a 'back-up' resource for people in the delta region. Should living in the metropolis become unsustainable as a result of climate change, pollution of the environment or an environmental disaster undermining the resilience of urban areas, people may choose to move away into new smaller cities.

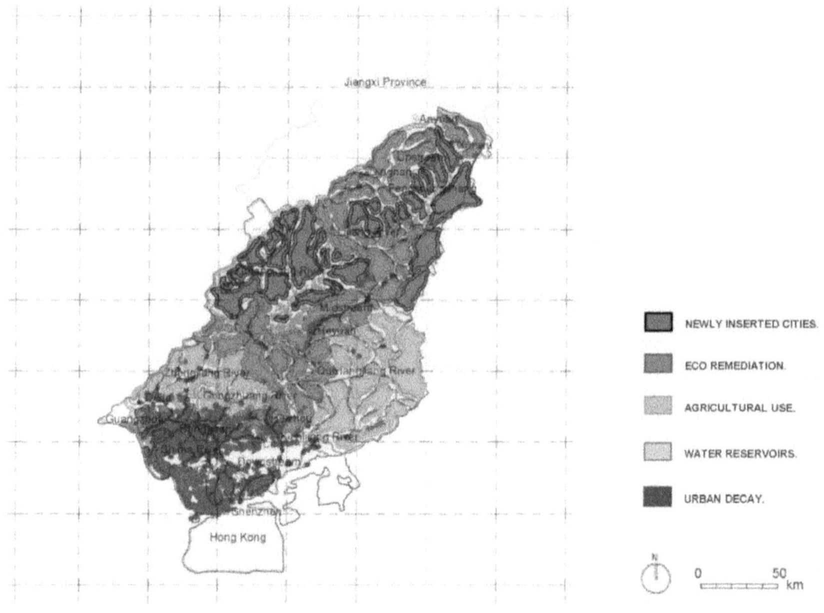

Fig. 3.40: The hypothetical scenario suggests that population will relocate from overcrowded city centers to smaller cities. New cities will be eco-friendly, people-oriented and have independent food production capacities. Image Source: by the author.

Flatness of arable land for farming and new settlements based on information on the elevation and slope of terrain.

This scenario considers the topography in the mountainous region as presented in the topographic elevation map (refer to Fig. 3.27) and identifies flat land in the mountain valleys suitable for agricultural activities. Land with a slope below 5° is considered as 'flat' and therefore suitable for agricultural cultivation.

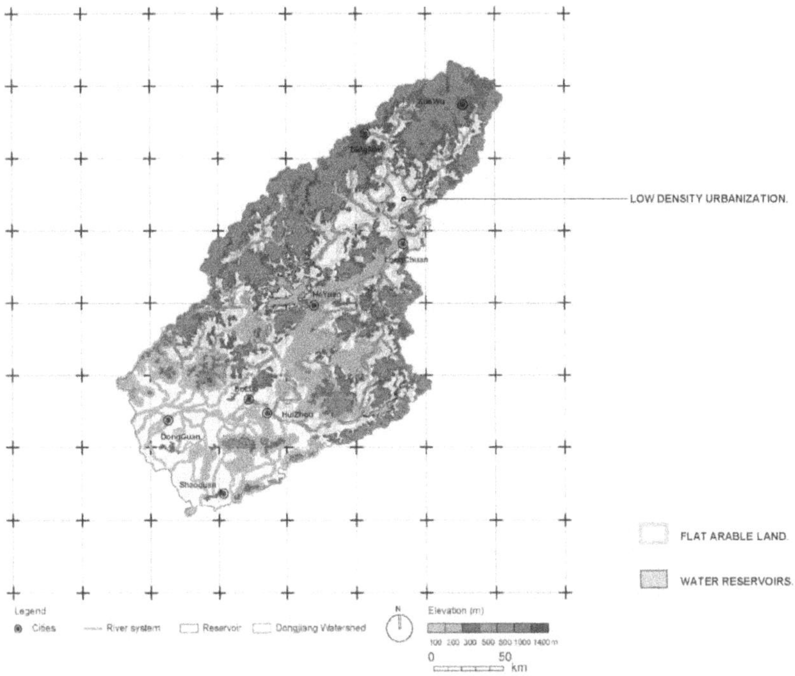

Fig. 3.41: Nestling of new settlements based on the topography. Image Source: by the author.

Low-density dispersed urbanization in the flat valley and preservation of forestry in steeper topography.

In this map, steeper topographies, with a slope above 10°, are designated for forestry. The land in the mountain valleys, with a slope below 5°, is chosen for the development of new low-density settlements to avoid higher costs of building works and transportation in mountain regions.

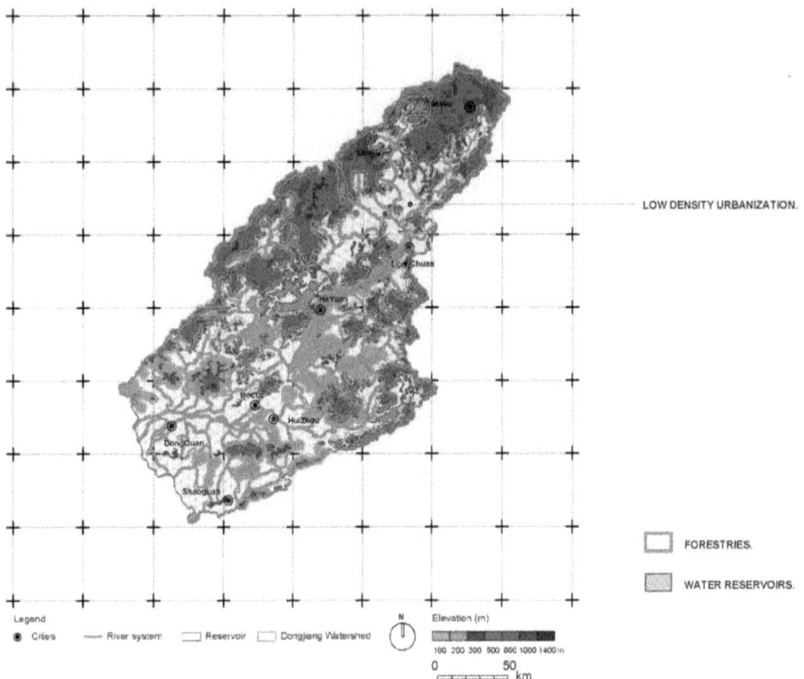

Fig. 3.42: Light red mesh shows dispersed communities in the flat valley of the topography. In this scenario the steeper mountain topographies are preserved for forestry. Image Source: by the author.

Tourism as a resource in mountain areas as an ecological and ethnic-cultural capital for the local economy and low-density communities development.

As a strategic plan, a combination of multiple interventions is considered for creating new communities in mountain regions. Tourism destinations are created to support the local economy. The tourism sites may specialize in ethnic-cultural tourism, which for this territory is the promotion of Hakka ethnic culture. Existing Hakka traditional buildings and temples are popular attractions for people in China and foreign visitors. Farming communities which specialize in agricultural tourism may set up fruit gardens and tea plantations for visitors from urban areas. Mountain scenic areas could also set up research facilities with linkages to universities in metropolitan areas. New mountain research centers may focus on biodiversity, environmental preservation and agricultural science. Local communities may further diversify into hybrid agricultural and industrial activities to evolve independent economic ecologies.

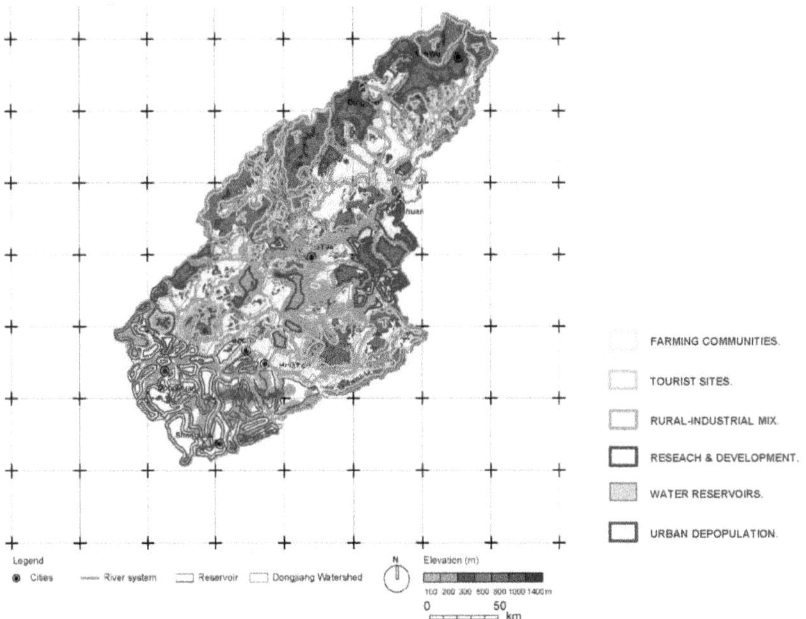

Fig. 3.43: Green outlines highlight landscape and scenic mountain regions with potential as tourist destinations. Farming communities are nestled in the valley regions in-between the mountains. Image Source: by the author.

3.2.2. Mountains to Mega-City analysis – a Transect approach.

Transect Studies and propositions.

The Mountain territory to Mega-city:

Dongjiang River Basin scale transect studies – Schematic.

Proposed Interventions in the Transects are elaborated in the following enlarged Transects see Fig.3.46 – 3.48.

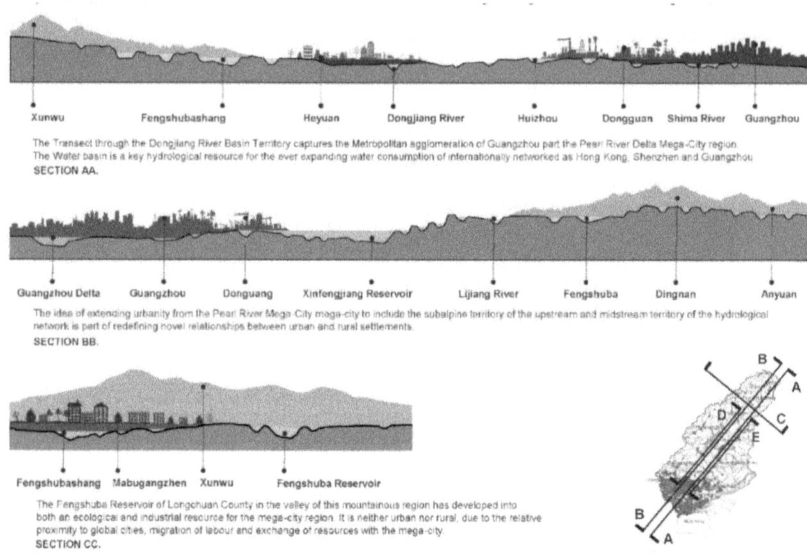

Fig. 3.44: Three main sections through the territories show the relationships from the Megacity to the mountains in the hinterland. Transect: by the author.

The Embedding of New Ecologies into the Territory:

Potential interventions (Part 1).

Fig. 3.45a: Part 1 of Table. Categorization of anthropogenic activities and potential new ecologies for the study sections. See potential interventions for a transformation shown in Fig. 3.46 - 3.48. Sources: Griboedov and by the author.

Potential interventions (Part 2).

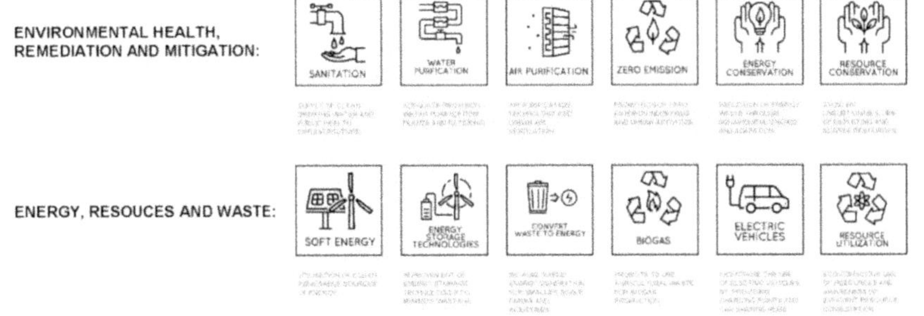

Fig. 3.45b: Part 2 of Table, see Fig. 3.45b. For the preparation of the potential interventions the objectives of the UN Sustainable Development Goals (SDG) 2015-2030 have been taken into consideration: https://sdgs.un.org/goals

Dongjiang River Basin Territory Scale Transect Studies – Schematic Scenarios.

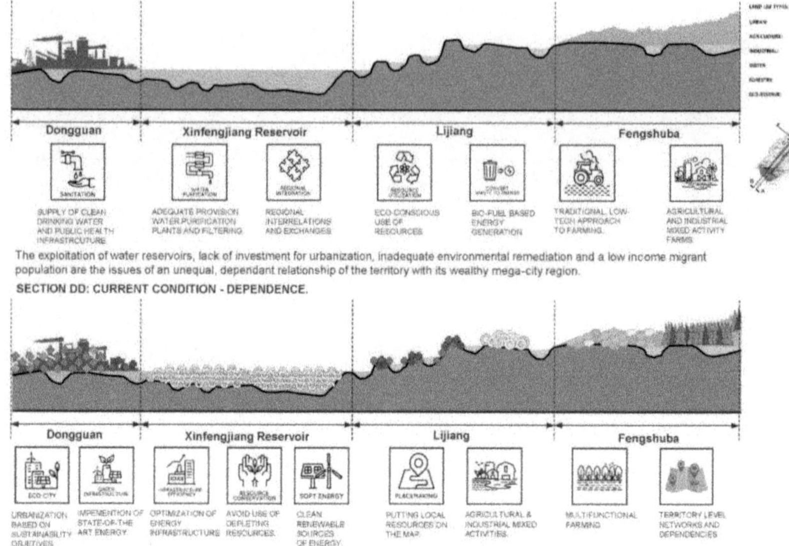

Fig. 3.46: The Upper section shows the existing ecologies the below sections proposes the embedding of new ecologies for greater independence and environmental remediation of the territory. Transect: by the author, pictogram: based on Griboedov.

Dongjiang River Basin Territory Scale Transect Studies – Schematic Scenarios.

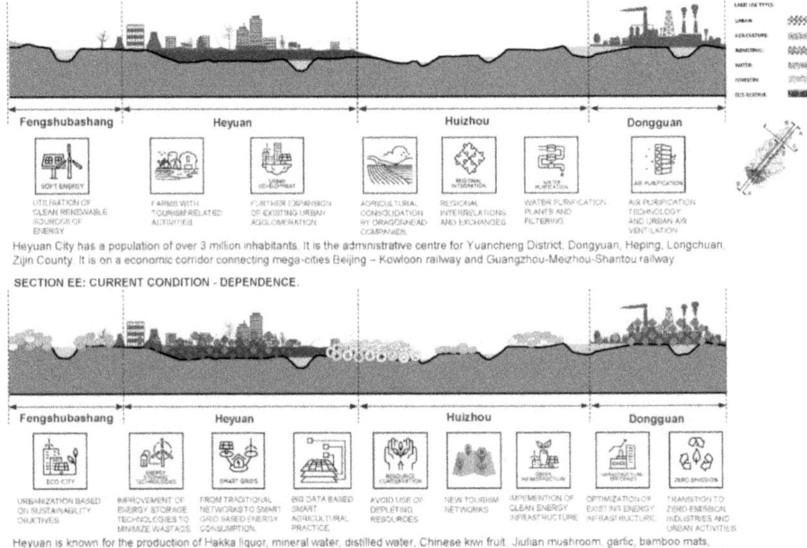

Fig. 3.47: refer to caption in Fig. 3.46. Transect: by the author, pictogram: based on Griboedov.

Dongjiang River Basin Territory Scale Transect Studies – Schematic Scenarios.

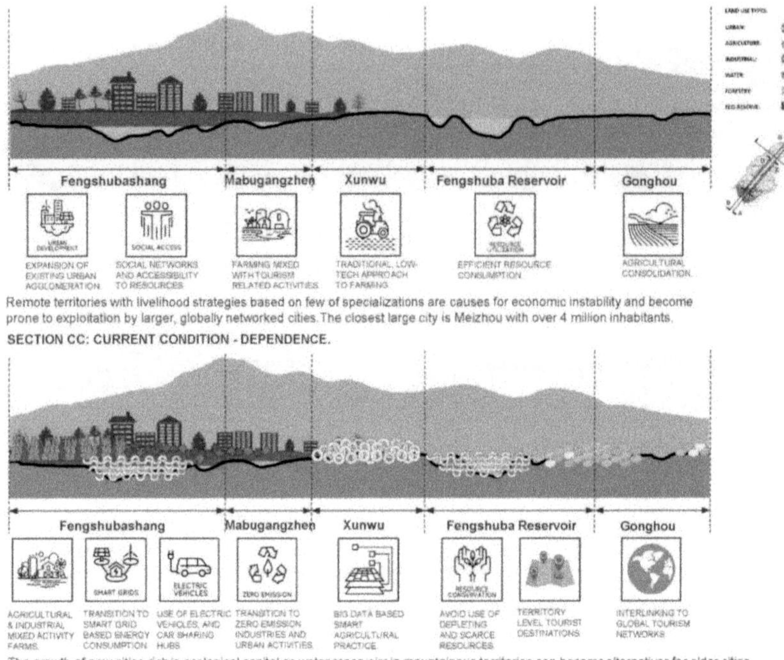

Fig. 3.48: refer to caption in Fig. 3.46. Transect: by the author, pictogram: based on Griboedov.

3.3. Scale of Investigation - 03.

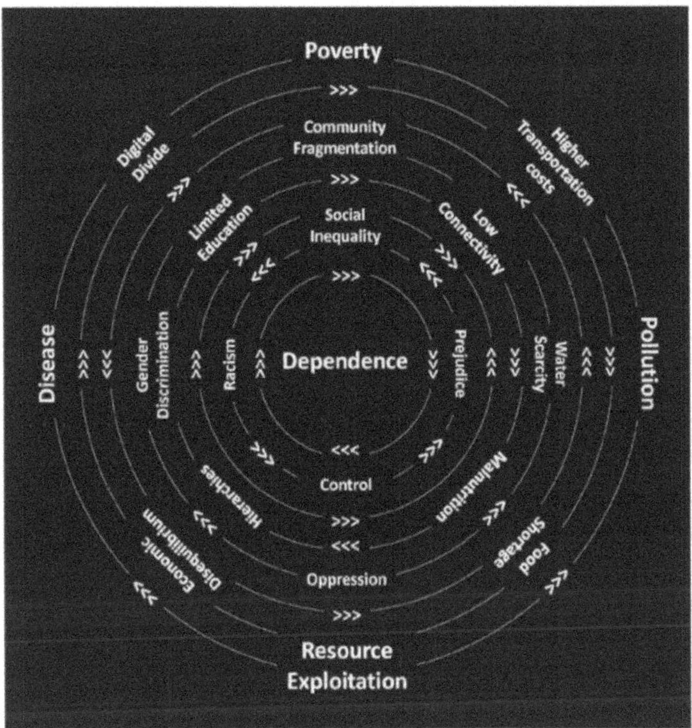

Fig. 3.49: The diagram illustrates a condition of Dependence for communities with its causes, associated negative externalities, social and environmental damages. Diagram: By the author.

The following section introduces, along a section line, three case study sites Chiguang town, Longbei town, and Changsha town, where field trips to the settlements have been made. Interviews with local residents were conducted to gain more detailed insights into the living conditions in the communities and livelihood strategies of local people.

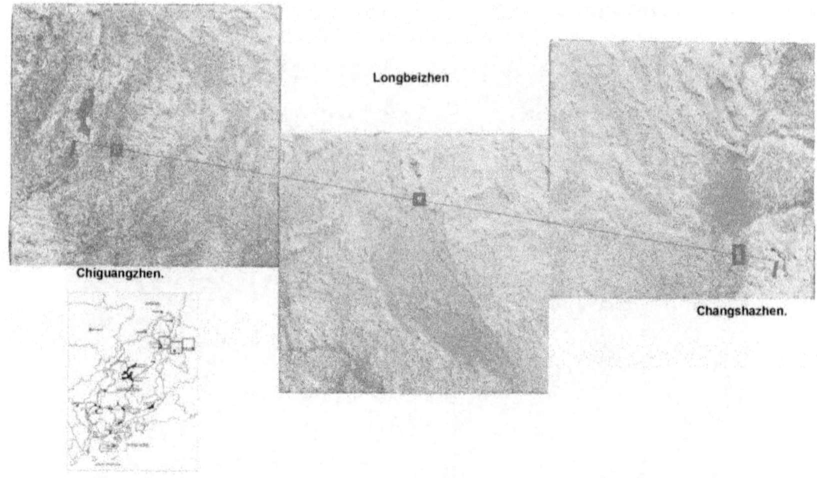

Fig. 3.50: Also see Fig. 3.85. Three study sites along a transect, Chiguang town close to Fengshuba reservoir, Longbei town at Heshui reservoir and Changsha town in Mei county. 3D-printed stereo-lithography of terrain: by the author.

3.3.1. Chiguang town - Fengshuba Reservoir.

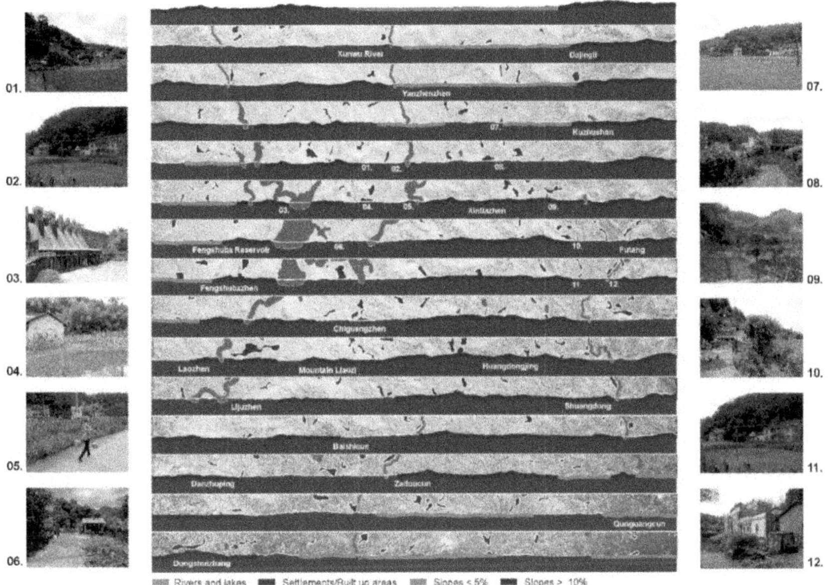

Fig. 3.51: Chiguangzhen town context with Fengshuba Reservoir: Sections through the topography overlaid on 3D-printed terrain: The sections in 10km intervals indicate the rivers and water reservoirs, with the map showing where settlements are located. Further, the sections identify the slopes < 5% (potentially suitable for farming activity, provided fertile soil is in place on site), and slopes > 10% are shown (not suitable for farming, unless terracing of topography is considered). Images: by the author.

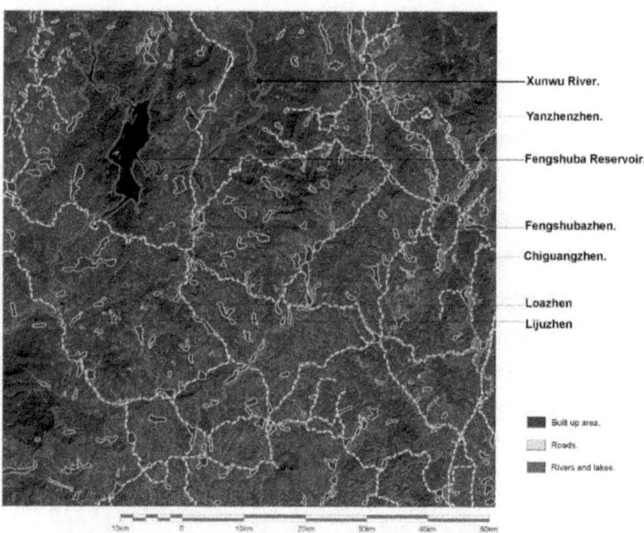

Fig. 3.52: Map of Chiguangzhen larger context with hydrological system, main roads and settlements shown. Sources: Google Earth & Image by the author with satellite map underlay.

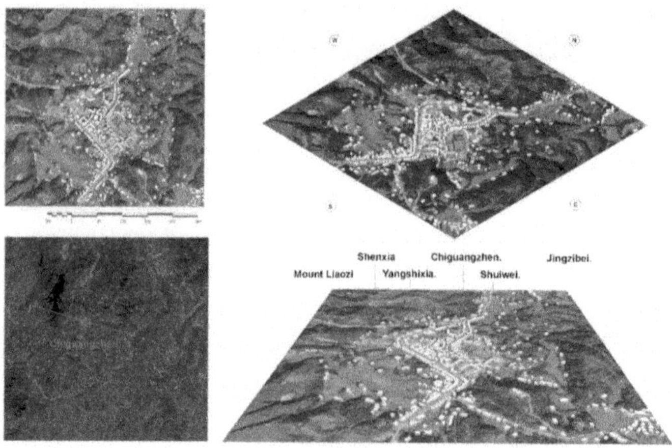

Fig. 3.53: The 3D visualizations show the settlement morphology of Chiguangzhen. The settlement is organized along a main road with in-line shops. Sources: Google Earth & Images by the author.

Fig. 3.54: Photographs F01 - F09: Fengshuba Reservoir dam and surrounding farms. Images by the author.

Fig. 3.55: Chiguangzhen, Photograph F10: Hakka courtyard houses are over one hundred years old. The spaces are open for community gatherings and for tourist visitors. Images: by the author.

Fig. 3.56: Chiguangzhen Town, Photographs F11 - F19: Images: by the author.

Fig. 3.57: Chiguangzhen, Photograph F20: Traditional Hakka courtyard house. The building is listed as a heritage building. Images: by the author.

Fig. 3.58: Chiguangzhen Town, Photographs F21 – F29: Images: by the author.

Fig. 3.59: Chiguangzhen, Photograph F30: Although Hakka courtyard houses are open to visitors; the buildings are still inhabited by local families. Image: by the author.

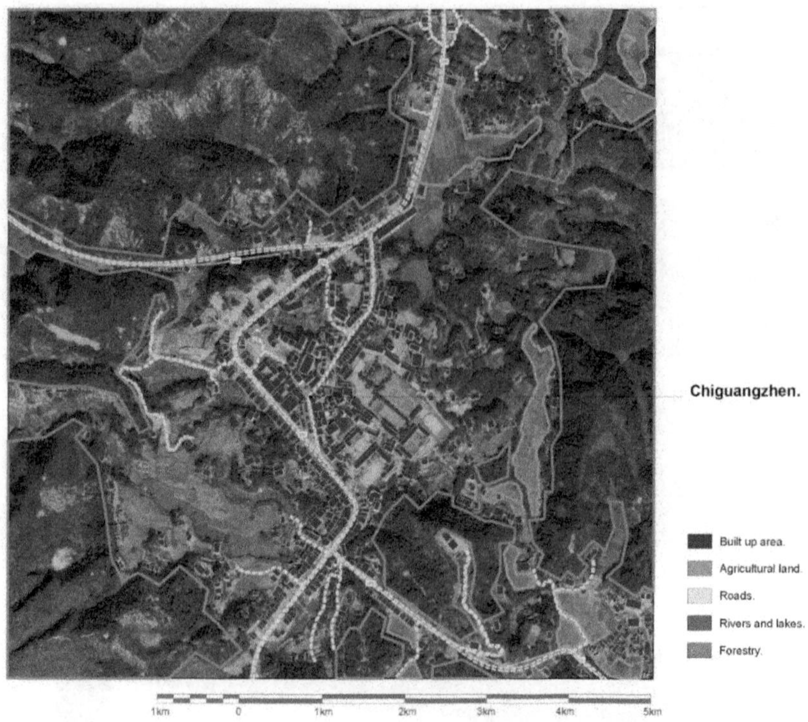

Fig. 3.60: Land use Map of Chiguangzhen town with built up area, agricultural land, roads, rivers, lakes and forestry shown. Sources: Google Earth & Image by the author with satellite map underlay.

3.3.1.1. Chiguang town: Existing condition – a Transect approach.

Key Issues in transect zone studied: Groundwater contamination, inadequate sanitation, illegal extraction of rare earth.

Poverty, Limited care for disabled people and the elderly. Shops & housing need renovation and upgrading.

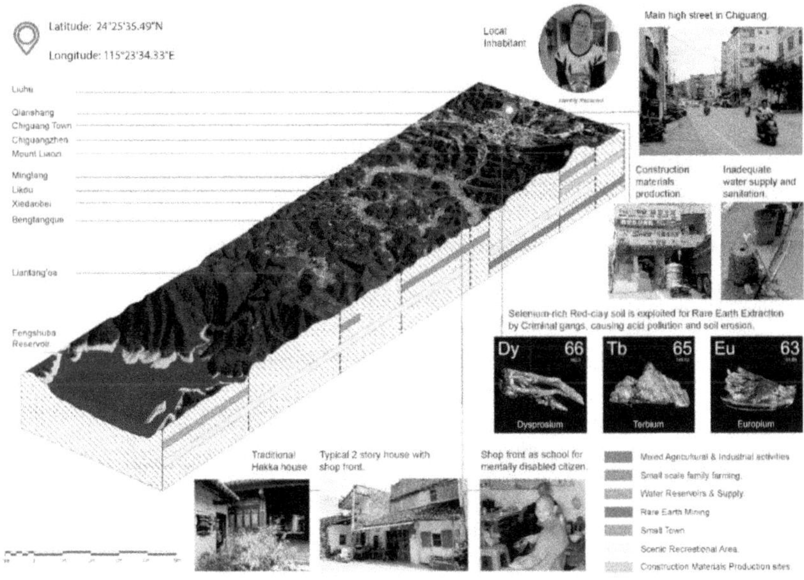

Fig. 3.61: Chiguangzhen Town with Fengshuba Reservoir is shown in axonometric section. The current livelihood activities are a mixture agricultural and construction material production. Illegal Rare Earth extractions of Dysprosium, Terbium and Europium are causing damage to agricultural land due to the acid washing of the soil. Ground water contamination is also an issue in this settlement. Rare earth extractions are sold to refineries for the production of electronic goods. Talking to local inhabitants, they noted the community needs better facilities

for disabled people. Currently, abandoned shops are used as schools for children with special needs. Images: Google Earth & by the author.

3.3.2. Longbei town – Heshui Reservoir.

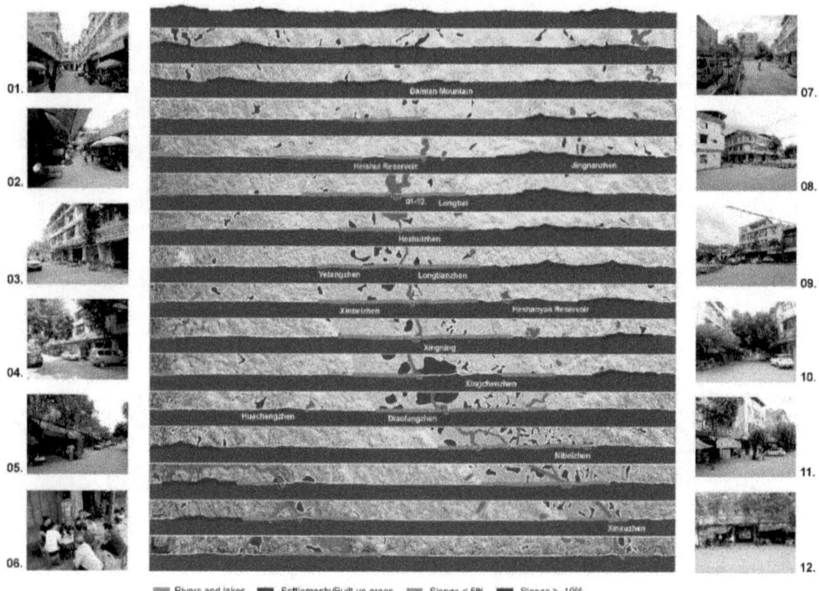

Fig. 3.62: Longbei town context with Heshui reservoir, and Xingning's dispersed settlement. Sections through the topography on 3D-printed terrain: Images: by the author.

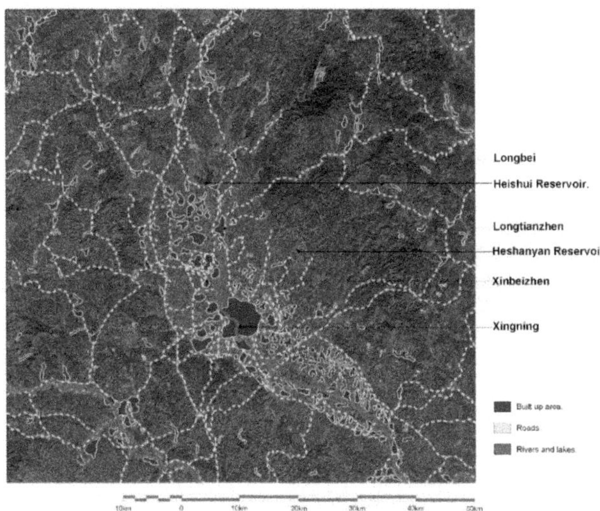

Fig. 3.63: Map of Longbei town larger context with hydrological system, main roads and settlements shown. Sources: Google Earth & Image by the author with satellite map underlay.

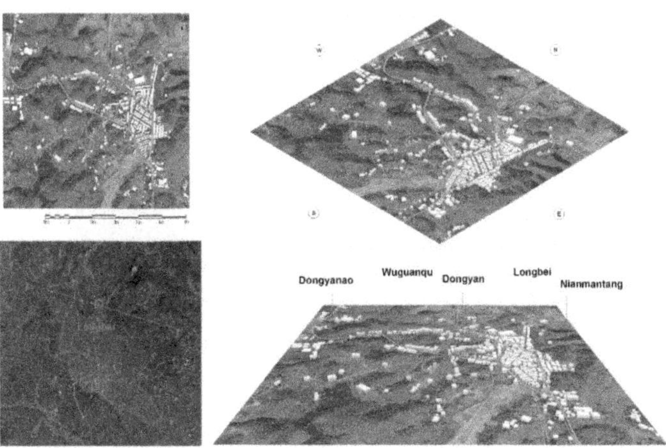

Fig. 3.64: The 3D visualizations show the morphology of Longbei. The town has in-line shop front streets connected with secondary back streets. Sources: Google Earth & Images by the author.

X01: Longbei street with in-line retail shops.
X02: Three-story high buildings with shop fronts.
X03: Grocery shop serving snacks and drinks.
X04: Taobao collection shop.
X05: The Taoboa shop has one PC workstation only.
X06: Local hair dresser.
X07: Fire station is used as community centre.
X08: People socializing to play Mahjong.
X09: Group of elderly playing cards at community centre.

Fig. 3.65: Longbei town, Photographs X01 – X09: Images: by the author.

Fig. 3.66: Longbei, Photograph X10: Longbei's wet markets require upgrading. Several of the shops are not in use or have inadequate sanitary facilities and electricity provisions. Images: by the author.

X11: Longbei's main square.

X12: Local in-line shopping street.

X13: Several of the shops are out of business.

X14: Local shoe shop with few goods on display.

X15: Shop with hardware goods.

X16: Pharmacy shop in main shopping street.

X17: Local people playing cards in front of shop.

X18: Children's shoes and clothing shop.

X19: Backstreet shops are largely not in use.

Fig. 3.67: Longbei town, Photographs X11 – X19: Images: by the author.

Fig. 3.68: Longbei, Photograph X20: Local vegetable shop attendant loading orders into a small van for delivery. Shops in the main retail street are in deplorable condition. Images: by the author.

Fig. 3.69: Longbei town, Photographs X21 – X29: Images: by the author.

Fig. 3.70: Longbei, Photograph X30: View of Longbei's streetscape. Most of the buildings are two story-high buildings. Taller town houses are up to four stories high. Images: by the author.

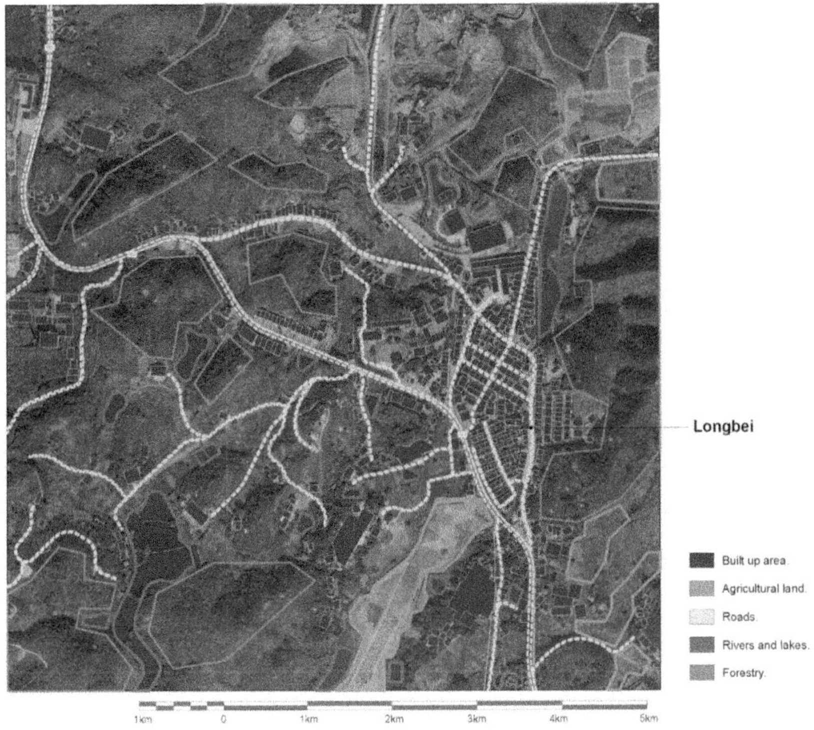

Fig. 3.71: Land use Map of Longbei town with built up area, agricultural land, roads, rivers, lakes and forestry shown. Sources: Google Earth & Image by the author with satellite map underlay.

3.3.2.1. Longbei town: Existing condition – a Transect approach.

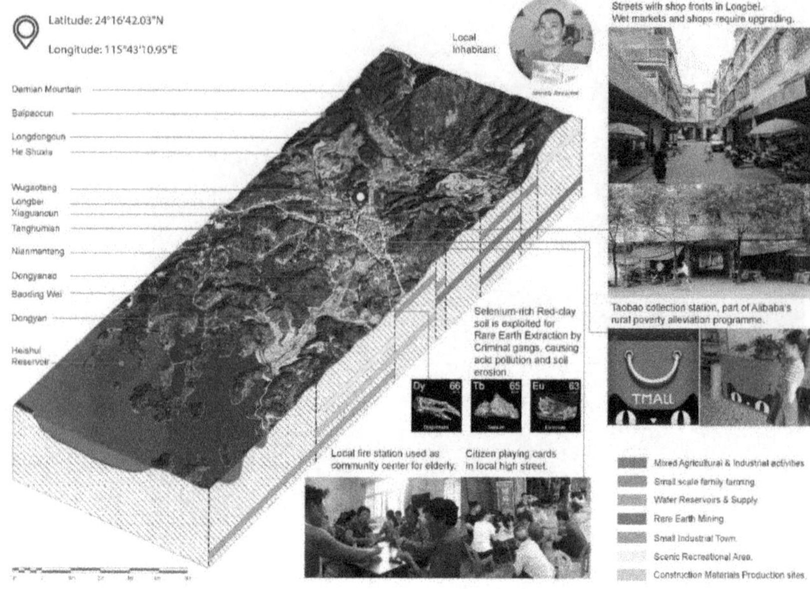

Fig. 3.72: Longbei Town with Heshui Reservoir is shown in axonometric section. In this town acute poverty and the pollution can be witnessed when walking through the main streets and visiting the shops. This settlement also has a basic Taobao distribution shop, which is part of the Alibaba poverty eradication program for disadvantaged communities. There was one employee and two customers when visiting the Taobao collection shop. The local wet market was empty during the visit, and in need of urgent upgrading. Illegal mining for rare earth is also practiced. The Heshui reservoir could be promoted as a tourist destination. However, raising external sources funds for setting up tourism facilities such as a water sports clubs and camping sites would need to the considered. The provincial government, 'private' corporations, charities or private donors could be approached for subventions. Images: Google Earth & by the author.

3.3.3. Changsha town - Meijiang River.

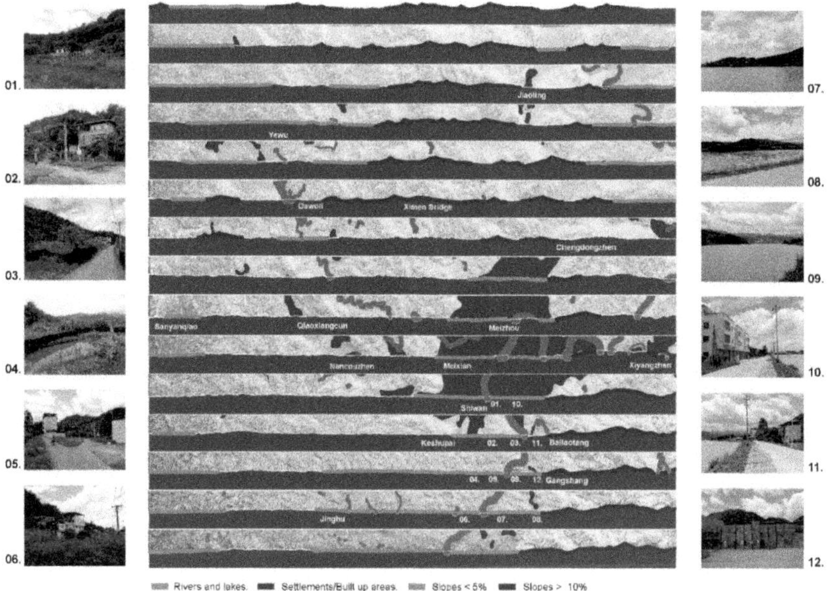

Fig. 3.73: Changshazhen town context with Meijiang River flowing through Meizhou: Sections through the topography overlaid on 3D-printed terrain: Images: by the author.

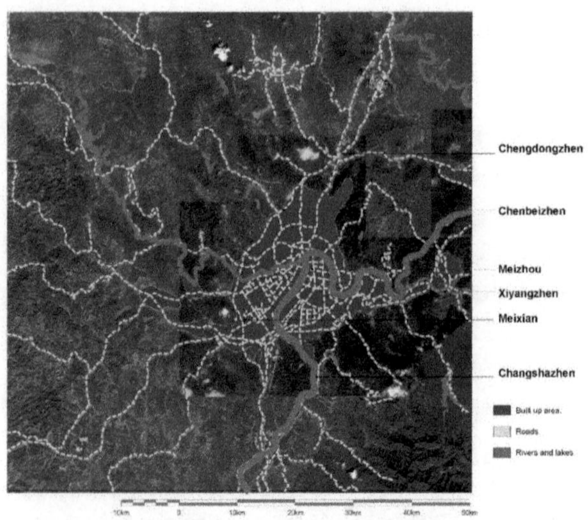

Fig. 3.74: Map of Changsha town larger context with hydrological system, main roads and settlements shown. Sources: Google Earth & Image by the author with satellite map underlay.

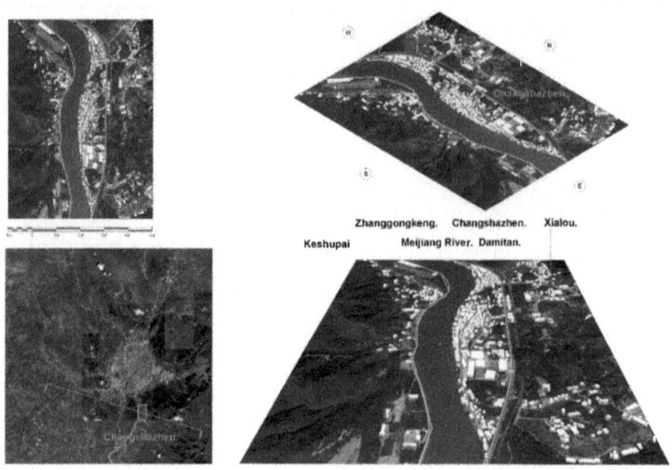

Fig. 3.75: The 3D visualizations show the settlement morphology of Changshazhen along the Meijiang River. The mixed agricultural and industrial activities characterize this settlement. Small

scale family farms typically have three-to-four story dwelling buildings for the older and younger generation of the family. The farming fields are surrounded by industrial sheds, construction material storage sites, new family houses and derelict traditional farms houses. Sources: Google Earth & Images by the author.

M01: View into Meijiang River with fisherman at bank.

M02: View showing industrial sheds of Changshazhen.

M03: Hilly parts with forests of the territory.

M04: Fisherman at river bank.

M05: Small scale mixed farming field.

M06: Three-to-four-story high family house types.

M07: Small buildings and sheds at farms.

M08: Family vegetable farm fields.

M09: Rice fields of small scale farmer.

Fig. 3.76: Changshazhen town, Photographs M01 – M09: Images: by the author.

Fig. 3.77: Changshazhen, Photograph M10: View along Meijiang River with semi-completed dwelling units and industrial sheds. Images: by the author.

M11: Potato fields of local farmer.
M12: Water reservoir of a family run farm.
M13: Small roads in between farms.
M14: Family homes of local farmers.
M15: Small water reservoir.
M16: Small neighbouring farm house.
M17: Lotus pond of a small farm.
M18: Traditional Hakka farm building.
M19: Road in-between fields

Fig. 3.78: Changshazhen town, Photographs M11 – M19: Images: by the author.

Fig. 3.79: Changshazhen, Photograph M20: View of scattered farm buildings to forests in Changshazhen. Images: by the author.

M21: View to factory buildings.
M22: Mixture of farm houses and industrial buildings.
M23: Motorcycles, a common low cost means for travel
M24: Typical three story high farm house.
M25: Newly built family house.
M26: Two semi-completed farm houses.
M27: Restaurant business in family house
M28: Motor bikes repair service.
M29: Local bricks and ceramic tiles plant.

Fig. 3.80: Changshazhen town, Photographs M21 – M29: Images: by the author.

Fig. 3.81: Changshazhen, Photograph M30: Traditional single-to-two story Hakka farm houses can be found in nestled in the landscape in-between rice fields and newer family houses. Images: by the author.

Fig. 3.82: Land use Map of Changshazhen town with built up area, agricultural land, roads, rivers, lakes and forestry shown. Sources: Google Earth & Image by the author with satellite map underlay.

3.3.3.1. Changsha town: Existing condition – a Transect approach.

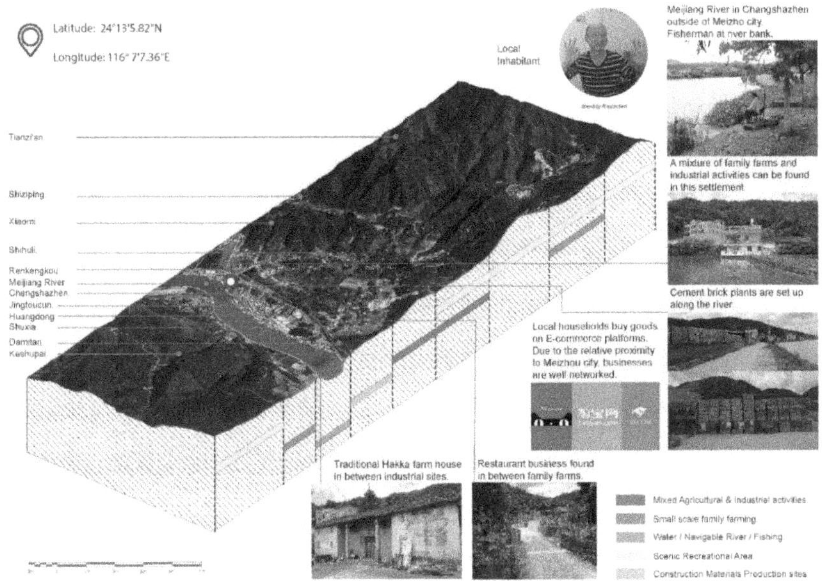

Fig. 3.83: Changshazhen in the Meijiang River Valley is shown in axonometric section. The small family farms and constructional material industrial sheds can be found along the river. The production plants specialize in making bricks and cement-based masonry blocks. Material storage set ups can be seen and delivery trucks area loaded with bricks for dispatch. The low-density settlement has some advantages, for example its proximity to Meizhou and the surrounding mountain scenery. The family farms could diversify into agricultural tourism. Few restaurants offering Hakka cuisine can be found. Access and connectivity for delivery of good by several Chinese e-commerce companies such as Alibaba and JD is set up. Images: Google Earth & by the author.

3.3.3.2. Yannafei Tea plantation example in Mei County.

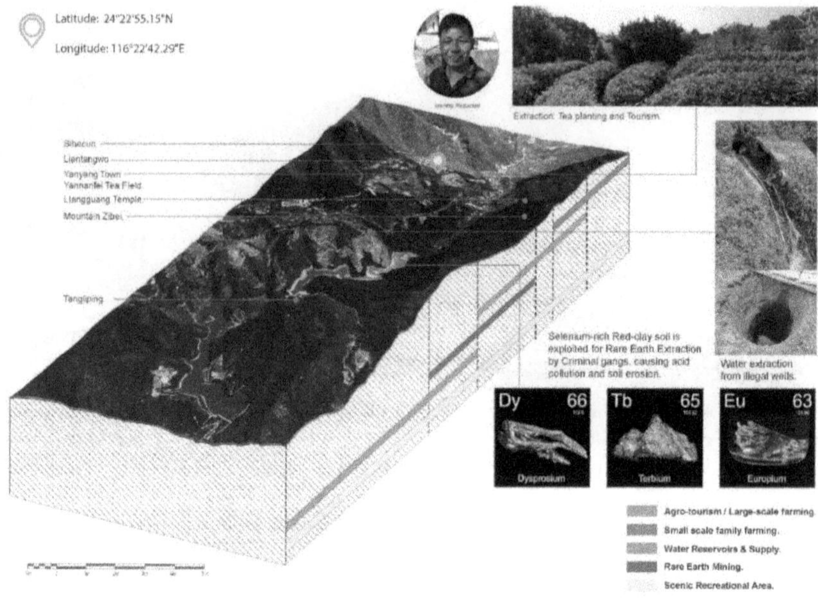

Fig. 3.84: The Yannafei Tea plantations in Mei country is a popular tourist destination attracting visitors mainly from Guangdong as well as neighbouring provinces in China. The tea plantations were visited as part of the fieldwork trips in the region. Talking to a local mountain dweller illegal digging of wells for water and extractions of metals is an issue for local farmers. Images: Google Earth & by the author.

3.4. Scale of Investigation - 04.

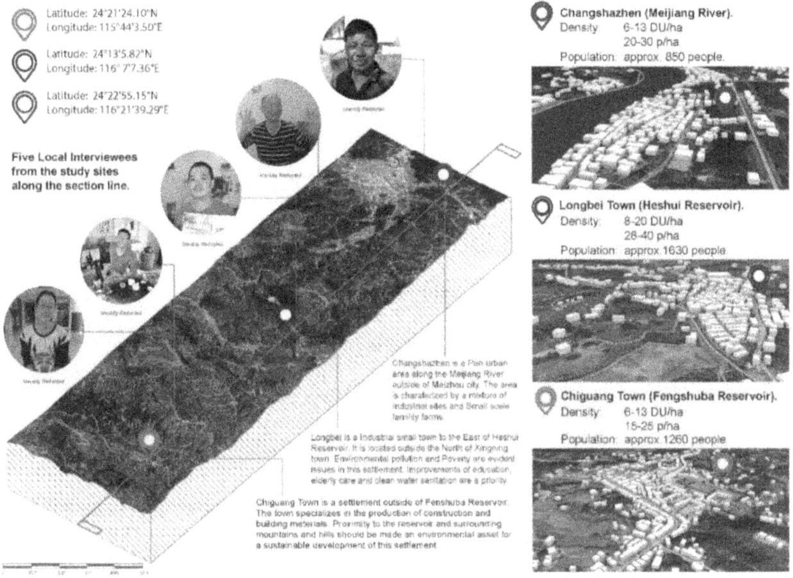

Fig. 3.85: Also see Fig. 3.50. Five Livelihood Scenarios of interviewees: Shown along a section through the Mei county territory in the context of Chiguangzhen, Longbeizhen, Changshazhen study sites. Images: Google Earth & by the author.

Local residents living within Mei County with different livelihood activities were visited in their communities.

Based on survey of in transects, peoples environments and the settlements they live in, the following scenarios construct grass-root level livelihood transformations. Alternatives to the idea that E-commerce and online retailing can lift people out of poverty are presented in Fig.3.89-3.93. Over time the livelihood innovations may assist people with economic independence and self-sufficiency. Communities with diversified livelihood strategies are could evolve into hybrid rural-urban settlements.

Fig. 3.86: Formal and Informal Interviews with local residents were conducted through a translator. The local inhabitants speak the Hakka dialect, linguistically related to Cantonese. Photography: By the author.

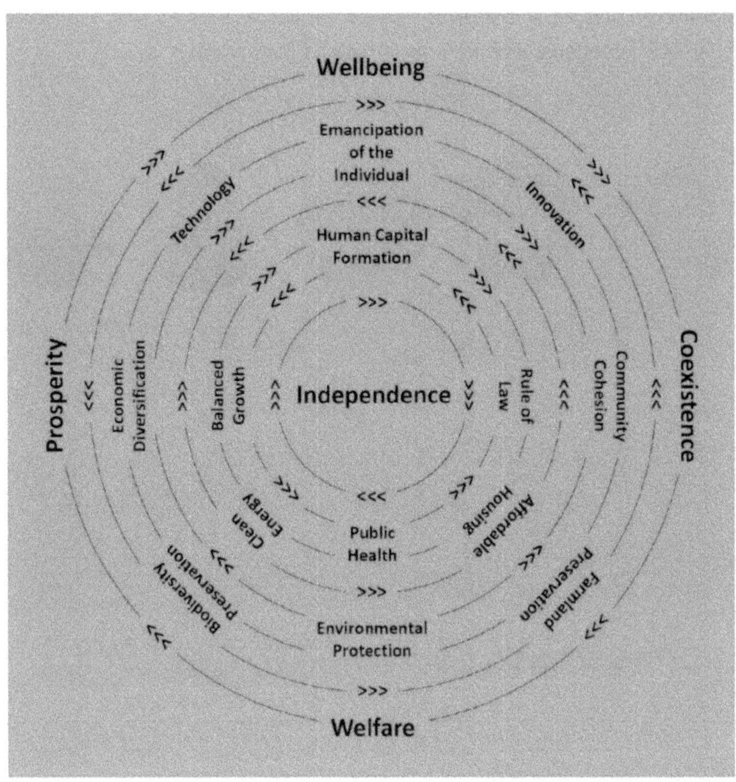

Fig. 3.87: 'Independence' is the ideal condition for remote settlements and the inhabitants. The diagram captures the key determinants and qualities needed for independent livelihoods.

Estimating the degree of livelihood independence for the Five Characters based on six areas of concerns affecting marginalized communities:

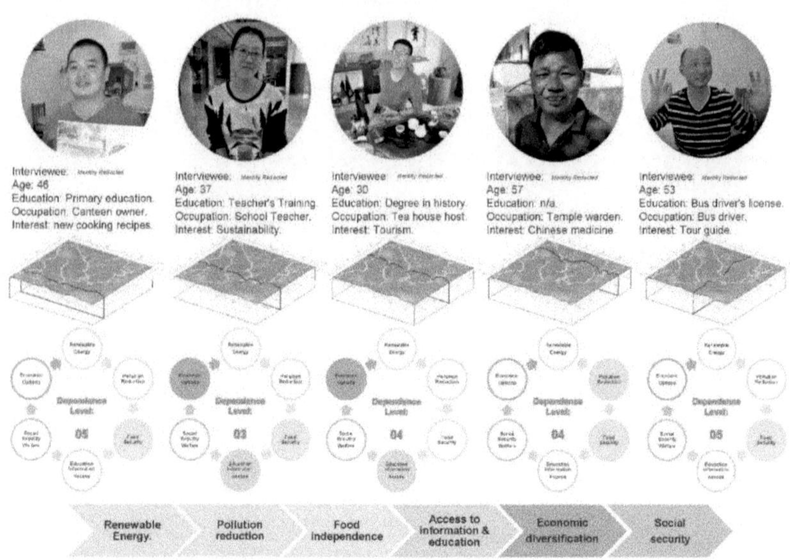

Fig. 3.88: While ideally, the five to six of the independence objectives would be met for a relatively balanced livelihood, at most the five local characters met 13 of the determinants. Therefore they have been grouped in higher Dependence Levels of 3 – 5. Source: by the author.

Character 01: Livelihood diversification strategy:

New Ecology Scenarios by the Transformation of Micro-Economies.

Case 01: Food – Organic Farming – Delivery services.

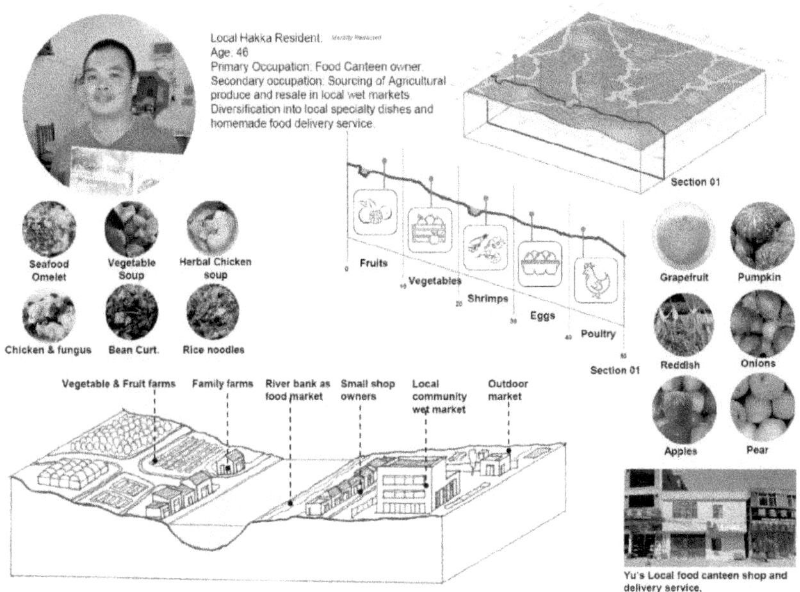

Fig. 3.89: Mr.Y. is a small canteen owner offering meals for local residents. He has enhanced his business by specializing in local signature meals. He sources all the produce through a local network of organic farmers. Additionally, he runs a delivery service for cooked meals and organic farming produce. Source: By the author.

Character 02: Livelihood diversification strategy:

Enhanced Livelihood Scenarios by Embedding of Micro-Ecologies.

Case 02: Off-grid Eco-lodge – Smart Agronomy – Sustainability Educator.

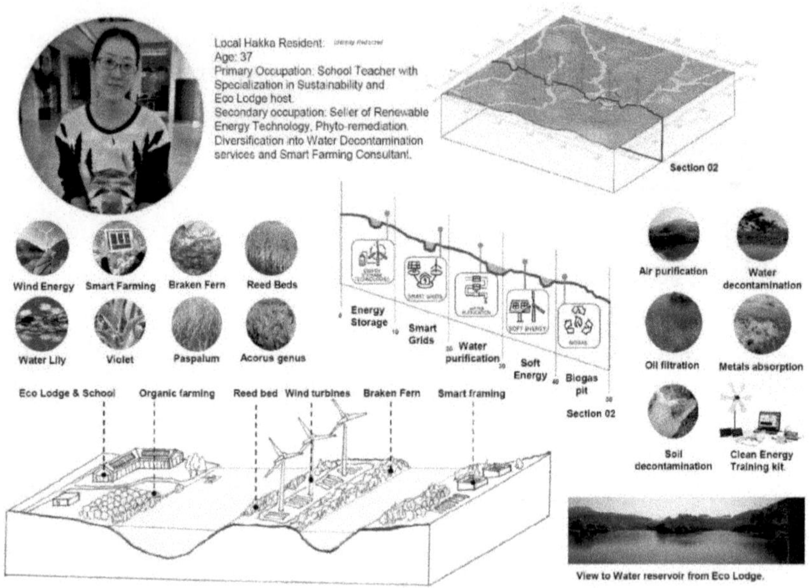

Fig. 3.90: Ms. H. set up a clean energy operated, off-grid Eco-lodge. She is also a smart farming consultant and Sustainability school teacher. Source: By the author.

Character 03: Livelihood diversification strategy:

New Ecology Scenarios with community involvement. Case 03: Guest & Tea house – Community centre - New Mobilities.

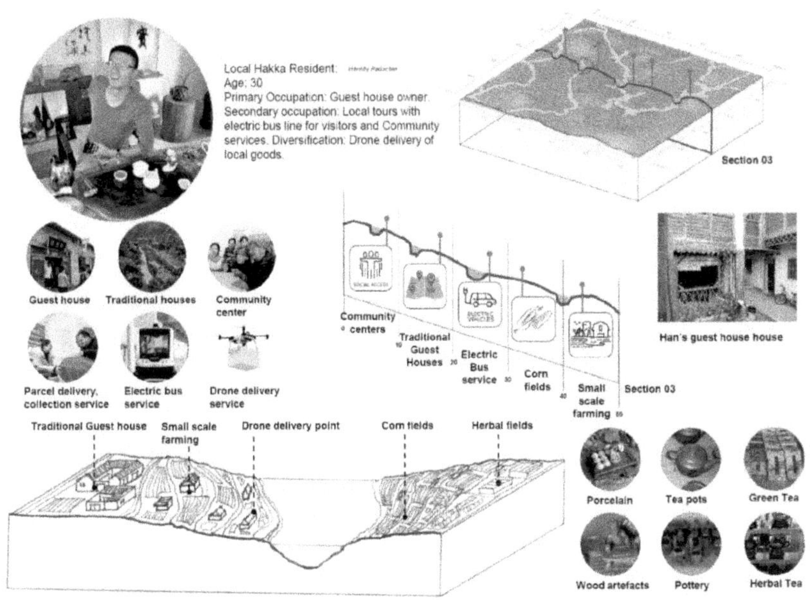

Fig. 3.91: Mr. H. is a guest house and tea house owner. He is also involved in community services. Additionally, he is a driver for electric mini buses service for elderly and has a drone pilot license. Source: By the author.

Character 04: Livelihood diversification strategy:

Livelihood Scenario for a mountain dweller.

Case 04: Temple Visitors guide – Chinese Medicine – Contract farming.

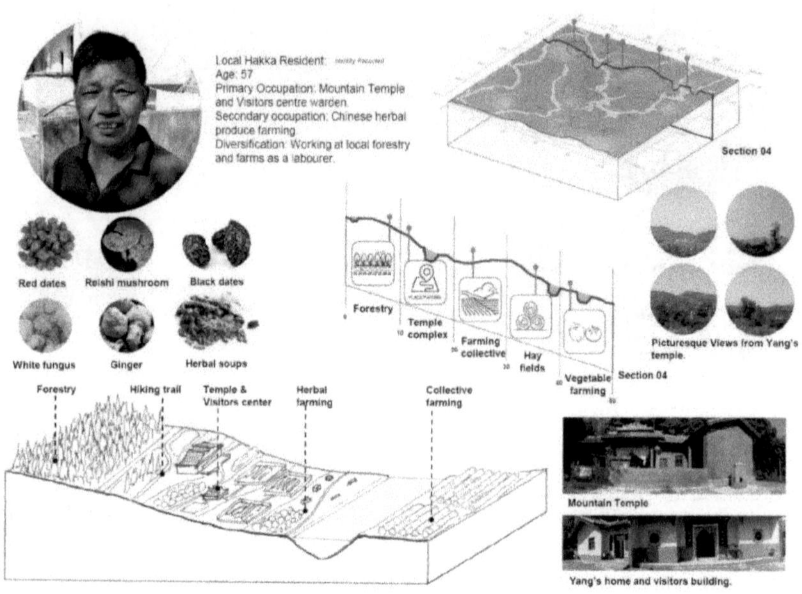

Fig. 3.92: Mr. Y. is a tour guide in a mountain temple complex. He grows and sells Chinese medicine produce. Occasionally, he also works as a contract farmer on local farming collectives. Source: By the author.

Character 05: Livelihood diversification strategy:

"Countryside Tourism" Scenario. Case 05: Agro-Tourism – Tour Guide – Bus & Courier driver.

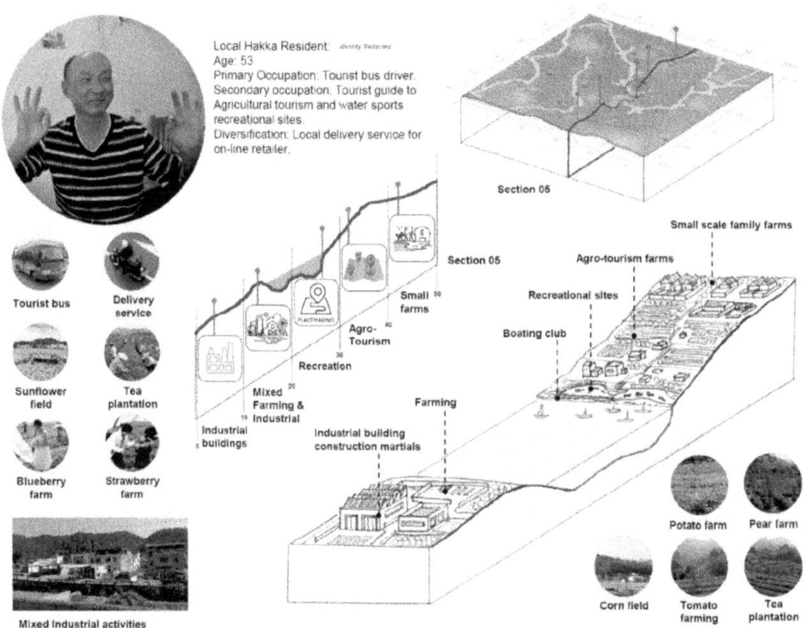

Fig. 3.93: Mr. F. is a Tourist Bus driver to local Agricultural Tourism farms. He also guides visitors to local water reservoir recreational sites. He also sells and delivers farming produce from the farms he visits.

3.5. Summary & Synthesis of design interventions:

The following design proposals for Chiguang town, Longbei town, Changsha town in sections 3.5 and 3.6 are the synthesizing part of the research by design studies in Part Three of this work. The following proposals suggest remedial interventions in the three distressed settlements studied. Refer to the sections with compensatory overlays shown in Fig. 3.97a -3.97c. To provide more detailed design paradigms of environmentally-friendly and welfare-oriented improvements for the communities, examples of street transformations in the Chiguang town, Longbei town, Changsha town have been developed. The street transformation designs have been visualized as perspective views and sections in Fig. 3.98a – 3.98i.

The design proposals are intended as an alternative of the rural development agendas proposed by Chinese E-commerce giants. A documentary video published by the LA Times (Makinen, 2016) addresses the rural expansion program by Alibaba, and the setting up of rural Taobao shop fronts in neglected territories (refer to Fig. 3.94.).

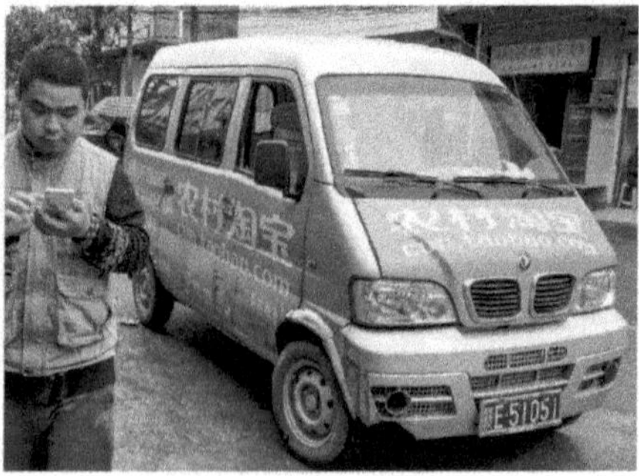

Fig. 3.94: Shop front of a Rural Taobao service center in Yufeng village (left). Taobao delivery van sourcing agricultural produce from farmers in Yufeng village for selling goods through Alibaba's online retailing platform (right). Image Source: Julie Makinen/ L.A. Times, 2016.(Documentary Video link:https://www.youtube.com/watch?v=1YueY9OXTF0&feature=emb_logo)

Fig. 3.95a: Alibaba Poverty Relief Fund announced by Alibaba's CEO Jack Ma in 2017 set up 10 billion yuan for poverty alleviation in rural and marginal mountain regions. The E-commerce giant pledged to address environmental remediation, rural entrepreneurship, to support women and to promote education. A long-term involvement and assistance by Alibaba executives with the sustainable development project could deliver long-lasting poverty eradication. Image Source: Xinhua 2017 / www.news.cn.

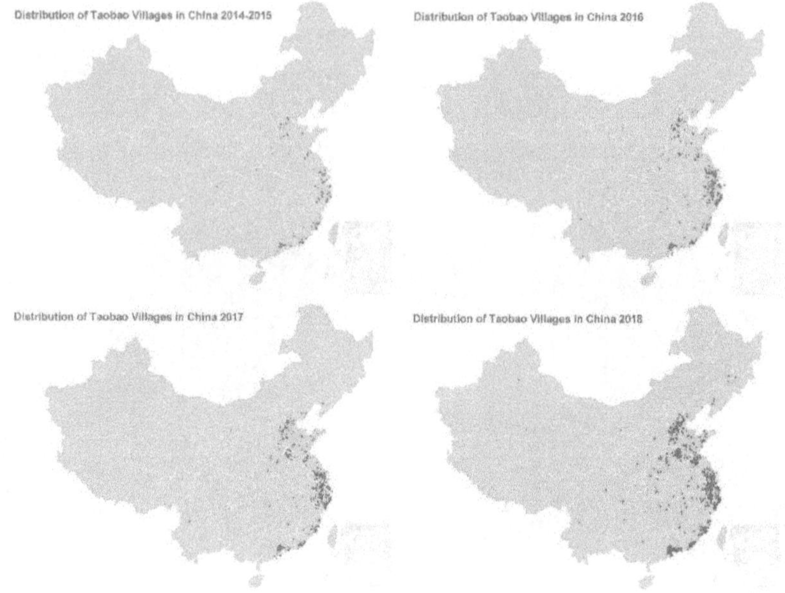

Fig. 3.95b: Distribution of Taobao Villages in China, 2014-2018. In 2019 Alibaba reported over 4000 Taobao Villages have been created. The maps by AliReseach show most of the e-commerce distribution centres in provinces along the coastal region of China. There are also some Taobao villages set up in the poverty- stricken, land-locked Central and Western provinces of China. Image Source: AliResearch, 2018.

The design studies in this research project propose alternative and diversified approaches to the idea that online retailing platforms and goods delivery networks alone underpin rural poverty alleviation. The expansion of e-commerce delivery networks into remote areas in China can contribute to the diversification of livelihood activities. However, as illustrated in Fig. 3.96e, it is regarded as only one of the possible strategies among several interventions needed. Equally relevant are the development of a green infrastructure and the provision of social welfare infrastructure for communities. As exemplified in the design proposals in the last section 3.6, green infrastructure would include the creation of additional parks, the planting of trees, increasing the vegetative coverage in the settlements, and renewable energy generation. As part of the measures for implementing social welfare functions in the disadvantaged settlements, the following designs suggest the recycling of existing, underutilized buildings into community walk-in clinics, sports and rehabilitation centers, kindergartens, mother & child care centers, a community library and a computer learning center.

In addition to preparing an environmental and welfare infrastructure, the lives of individuals, their determination and prospects for a better future are as important as their ability to build new economic ecologies. For the emancipation of a community as part of a sustainable development project it is equally essential for the citizen to rethink daily life (Knierbein & Viderman, 2018) at grassroots level. Community participation instead of inertia, and the engagement of individual citizens seeking improvements to existing living conditions in settlements plays a crucial role. Existential global issues including the environmental crisis, poverty, water shortage and food insecurity also have an impact on the quality of life in remote rural areas. In particular, socially and economically vulnerable people and their families are most affected by competition for resources. Hence, the support and a social welfare system for dependent individuals in a community are necessary for forming self-sufficient communities.

The inevitability to reverse the process of global urbanization and to further develop rural areas calls for independence planning as part of regional design projects. As exemplified in the transect design studies, the embedding of new livelihood activities as part of the design process in sections is suggested as a trajectory to evolve an alternative urbanity.

In the case of fragmented rural-urban territories in China, 'Acupunctural interventions' in transects are an incremental planning method explored in the design studies of this research project. In other words, transects help us to understand underlying natural environmental and local livelihood ecologies in territories. However, assuming that continuities in eco-systems function without disruptions can no longer be taken for granted. On the contrary, eco-system stabilities are disrupted and fractured, due to human intervention in the environment. Additionally, digitization & new geographies have increasingly altered human livelihood activities and their transactions in space and time. Therefore, remediation of fragmented eco-system services is urgent with transects as a design tool. Additionally, localized inclusions and overlays of recent technological innovations are becoming not only indispensible, but also integral to sustainable settlement development. In contrast, the New Urbanist Movement promotes a rural-to-urban development, this highly idealized paradigms is more difficult to implement in the hinterland of Chinese Mega-cities, where uncoordinated industrial and agricultural hybrid uses of land have damaged the natural environment.

Potential Positive and Negative Externalities of Taobao Villages:

- Potential for incubator for grass roots start up in remote areas.
- Potential for new distribution channels.
- Potential for specialist/unique produce selling.
- Potential for supply and value chain ecologies.
- Potential for income diversification.
- Potential of ecommerce for poverty reduction in remote area.
- Potential employment for mothers & elderly.

- Oriented towards cheapest products / price wars.
- Products made in informal economies / urban villages & sweat shops.
- Dangerous working conditions / Child labor.
- Limited tax revenue/ limited regulation.
- Government expenditure for dealing with negative externalities higher than tax revenue.
- People in mountain areas often have poor education, low skills, limited access to transportation.
- Small orders for unique products / lack of economies of scale.

Source: Friedrich-Ebert-Stiftung - Taobao Villages, Fan Lulu, 2019.

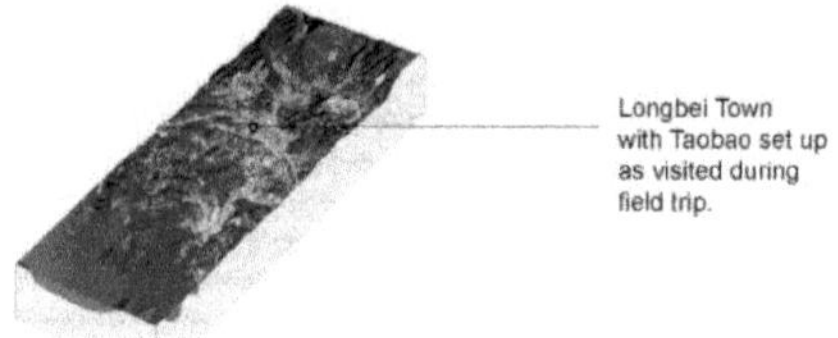

Fig. 3.95c: A Taobao distribution centre was visited during one of the fieldwork trips to Longbei town close to Heshui Reservoir. The shop is equipped with only one PC workstation and flat bed monitor, and a small counter for receiving customers. The set up of the shop with shelves for delivery goods is basic. Sources: Lulu, F.(2019),Taobao Villages - The Emergence of a New Pattern of Rural Ecommerce in China and its Social Implications, Retrieved 16.09.2019, https://www.fes-indonesia.org/e/en-taobao-villages-the-emergence-of-a-new-pattern-of-rural-

ecommerce-in-china-and-its-social-implications/; Axonometric and photograph generated by the author.

Perspective view into one of the valleys visited close to Yannafei Tea plantations (see Fig.2.18 and Fig.3.93): While agricultural-tourism supports the plantations with revenue from visitors, local farmers in the valley struggle with illegal natural resource extractions: The paradox of Rare Earth illegal mining and Poverty Alleviation programs promoted by Hi-Tech corporations.

Fig. 3.95d: The same e-commerce giants that claim to commit to poverty reduction, provide the on-line platforms for selling cheap smart phones with touch screens produced with rare earth extractions from impoverished rural and mountain areas. Extraction of rare earth for electronic products are often mined by criminal gangs in China, entering farms and illegal digging out metals and acid washing soil intended for agricultural cultivation. Image Source: Handsetrecycle.

By embedding self-sufficient economic activities, low density settlements which avoid an overdependence on larger cities, and can resists the exploitation of resources by a metropolis may grow into an independent city. An eco-friendly and socially-conscious community growth, as an alternative type city, abandoning a 'business as usual' pattern of development, would be based on a different set of values than under a capitalist form of urbanization. Its community leadership would prioritize environmental protection, health and sanitation, human capital formation and the production of culture.

As the urbanization of rural areas is still regarded as a solution for maintaining economic growth in China, a change in the priorities of community leaders will require more time to mature. The impact of growing the number of medium and small size cities in China in the near future will negatively affect the environment, resource consumption and the burdening of existing infrastructural networks. While China is transitioning into a phase of economic retrenchment, it is likely, that unintended, negative consequences of new urban construction impacting people, biodiversity and the environment will cause further damage in the next decade. Therefore, the development of small and medium sized cities, which take into consideration not only economic productivity but also environmental remediation, the well-being and social welfare for people, and respect local ethnic cultures would be a more evolved, next stage of urbanization.

China's National New Type Urbanization Plan (NTUP) for 2014-2020, a well intentioned policy can form a basic set of principles for a better phase of urbanization. The NTUP objectives (see Fig. 3.96a) includes a balanced urban-rural development, coordinated regional development, environmentally friendly and sustainable urbanization, biodiversity preservation, food security, poverty alleviation, and the preservation of ethnic identities. For the research by design studies in this research project, the following priorities in line with the NTUP have been emphasized: Renewable energy generation, pollution reduction, access to education, food production at small-scale farm level, welfare safety net building at community level (see Fig. 3.96d), economic diversification and Hakka ethnic cultural tourism promotion. Due to the relatively short period of time given by the official National New Type Urbanization Plan for 2014-2020, an alternative planning strategy considering a long-term impact recommends the ongoing updating of the development priorities and innovation of a Repertoire of Alternative Interventions (see as examples Fig. 3.96b & 3.96c).

This principle of maintaining alternative possibilities (Frankfurt, 2014) applies to all scales of spatial planning: the individual family farm scale, village community scale, small, medium, large cities and the regional scale. Thus, it is understandable that transforming the urbanization objectives and abandoning unsustainable development patterns requires both bottom up civic engagement and top down sustainable development policy implementation. Hence for reinstating ecological and community stability in mountain regions a time span until at least 2050 gives a better prospect for forming people-oriented cities.

Further, two important additional aims for a new urbanization phase are the implementation of NTUP objectives, not only as a high-level policy document, but also as concrete regional design projects by planning professionals, and the livelihood activities in co-produced settlements by its people and the community administrations. Similarly, the ideal condition of independence and emancipation of peripheral settlements can only be achieved by community participation, synergies between people (see Fig. 3.96f) and the protection of vulnerable members of society. Changes in livelihood choices beginning at grass-roots level initiatives of individuals and smaller groups of people must not be underestimated in terms of the positive impact it can have on the entire community.

However, the up-scaling of transformative initiatives at community level will depend on the organization of people into mutually supporting networks working towards the common goal of everyday life improvements and economic betterment. The consolidation of efforts with other mountain community networks can reinforce the scale of endeavors, social cohesion, and protection of common interests. Two examples of umbrella-organizations for marginal settlements from Europe are the CIPRA International in Lichtenstein which promotes sustainable development in Alpine regions, or the SIMRA (Social Innovation in Marginalized Rural Areas) an organization funded by the EU. Similar self-supportive mountain community networks could be formed in China in the future.

For achieving a level playing field with neighbouring settlements communities which may have reached more mature stages of urban development, the advancement of human capital formation, and ongoing protection of shared ecological capital are an indispensable shared interest. Consequently, an independently developing mountain community will be able to self-perpetuate its local economy, and simultaneously engage in economic transactions with other cities.

As the issue of emancipation in spatial planning is attracting attention from architects and urban planners, it can be anticipated that, the inclusion of Emancipatory Thinking in urban projects will gain momentum in the future. Given the challenges imposed by global economic and environmental struggles in marginal regions, its populations will likely protest against or reject the causes of spatial injustice and economic deprivation in peripheral communities. In particular, in China were a new, wealthier Chinese middle class is emerging, apart from capital and material accumulation, the wellbeing, intellectual growth and the possibility to live in ecology-friendly urban environments will likely be an aspiration for an increasingly educated and modern civilization.

The resettlement of the 'illegal rural' citizens in cities into new towns is also an urban and planning policy challenge requiring good design and planning proposals. Safe, dignified environments for working, living and environmentally friendly city design for a sustainable development of next generation communities will be priority.

With a re-orientation away from the concentration of development in the coastal cities, allocating resources to the rejuvenation of mountain settlements could be supported by the transfer of technological knowledge into the mountains. With the possibilities offered by recent developments (see Fig. 3.96e) in agronomy, new systems of mobility and communication, the transformation of landlocked mountain territories into an alternative urbanity could materialize in the future.

The poverty alleviation programs pursued by E-commerce giants (see Fig. 3.95c & 3.95d) in China can only be effective, provided that these initiative do not only serve the purpose of expanding an market-share of on-line consumers into countryside. A genuine concern for social and environmental issues, will not be solved by setting up parcel distribution centres equipped with internet, a computer workstation (see Fig. 3.95a). Investment by Chinese tech-companies into schools, hospitals, community centres, clean energy and water purification plants in communities would be more instrumental for poverty alleviation. In keeping with the idea of maintaining both independence of mountain cities and the forming of new synergies with neighbouring settlements, an ultimate end result may be a fair and just integration of mountain territories into megacity clusters. In this way a distinction between 'centre' and 'periphery' may be dissolved.

Considering an Emancipatory Urbanization Path for 2020-2050+

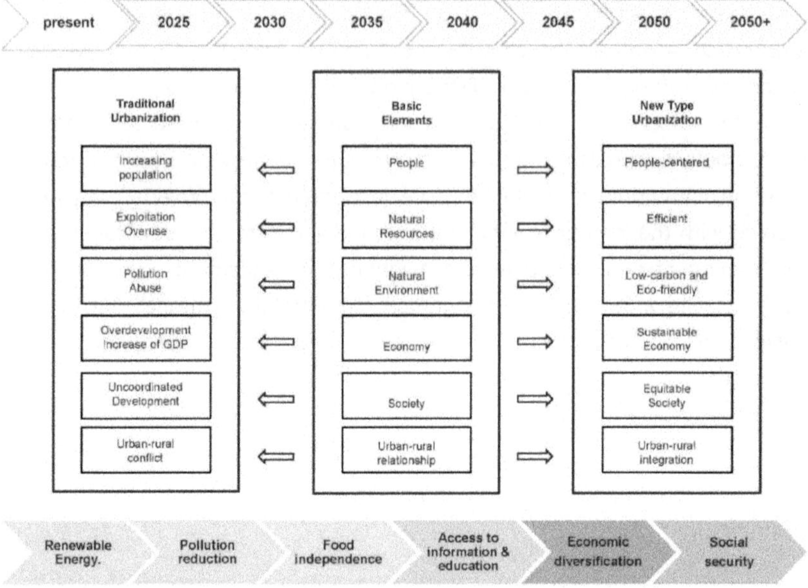

Fig. 3.96a: As a further development of the New Type Urbanization Plan beyond 2020, the following focus areas have been identified for the research by design studies for the Present to 2050+. The development objectives have selected for the chosen study sites to best achieve emancipation and to become more economically and resource independent settlements: (1) Renewable Energy Independence, (2) Pollution Reduction, (3) Food Independence, (4) Access to Information and Education, (5) Economic Diversification, (6) Social Security. Sources: Xu, Wang, Zhou, Wang, Liu, 2016; and by the author.

Potential Repertoire of Interventions: Remedial measures & livelihood strategies. Possible development stages for a new type of Urbanity: present to 2050+

Due to the interventions at grassroots level a longer time frame is anticipated for the maturing of emancipated urbanities

The indicative phasing of forty-two strategic measures for an Independence scenario & resilient livelihoods.

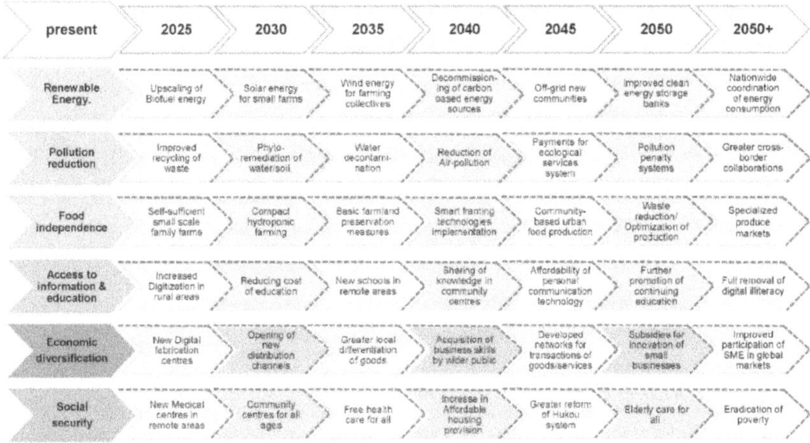

Fig. 3.96b: The timeline diagram for the present to 2050+ proposes incremental local interventions for the making of local ecologies and an emancipated urbanity. Neither the sequence nor the type of interventions are prescriptive. The main purpose for planning for the next phases of urbanization is to have repertoires of possibilities prepared in times of uncertainly. Source: By the author.

Principle of Alternative possibilities for regional planning.

Fig. 3.96c: principle of Alternative possibilities. The table lists potential interventions in distressed territories. Here table is provided for purpose of summary, for larger view see Fig. 3.45 a & b.

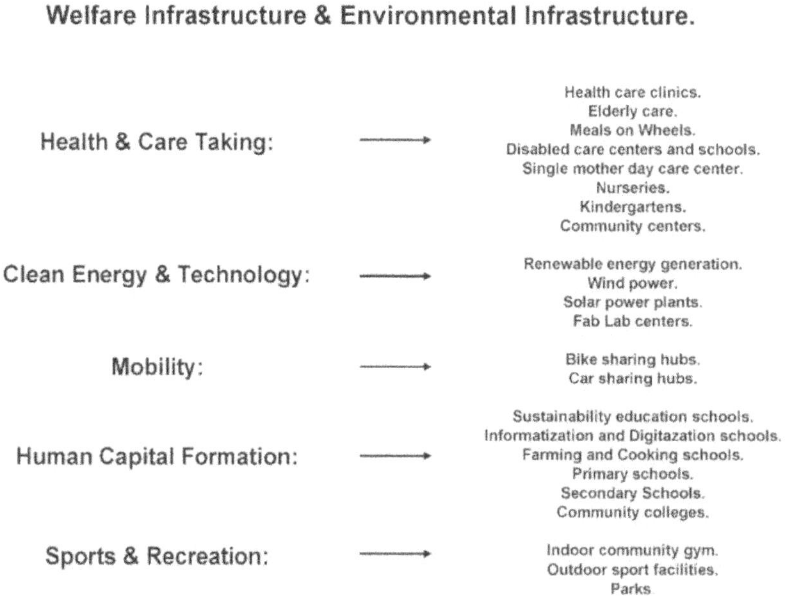

Welfare Infrastructure & Environmental Infrastructure.

Health & Care Taking:	⟶	Health care clinics. Elderly care. Meals on Wheels. Disabled care centers and schools. Single mother day care center. Nurseries. Kindergartens. Community centers.
Clean Energy & Technology:	⟶	Renewable energy generation. Wind power. Solar power plants. Fab Lab centers.
Mobility:	⟶	Bike sharing hubs. Car sharing hubs.
Human Capital Formation:	⟶	Sustainability education schools. Informatization and Digitazation schools. Farming and Cooking schools. Primary schools. Secondary Schools. Community colleges.
Sports & Recreation:	⟶	Indoor community gym. Outdoor sport facilities. Parks.

Fig. 3.96d: Giving support to marginalized communities on a path to sustainable development entails the provision social and welfare services which are part of a planning process. In particular, it considers vulnerable members of the community including please with health issues, children, mothers, disabled people and the elderly. Clean energy and energy independence is a shared responsibility for settlements in a process of adaptation for climate change.

Determinants of Livelihood diversification and Micro-Ecologies:

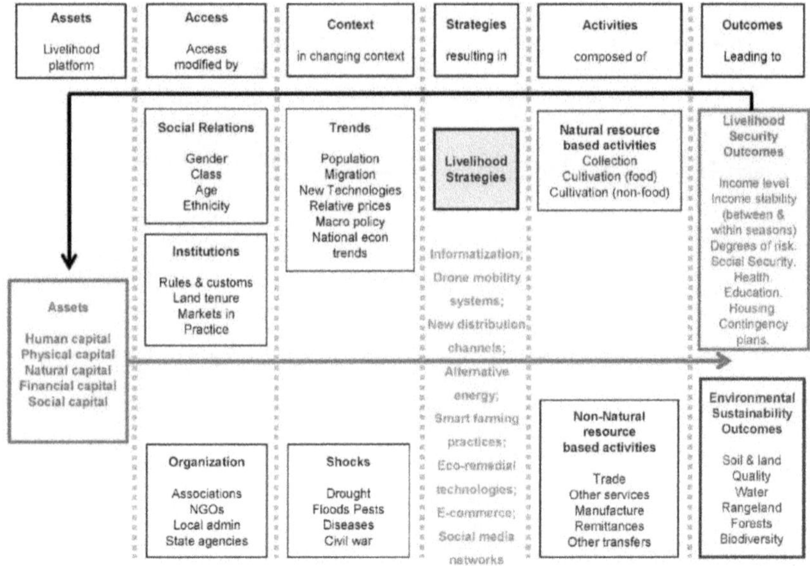

Fig. 3.96e: This table of livelihood determinants shows inclusion of recent technologies and know-how as part of an innovation and diversification strategy. The strategies include Informatization of remote territories, Drone mobility systems, opening of new distribution channels, alternative clean energy, Smart farming practices, Eco-remedial technologies, E-commerce, and Social media networks. Image Source: by the author, adapted based on Ellis, 2000.

Transformational processes by the multiplication of micro-economies and transactions between people in communities.

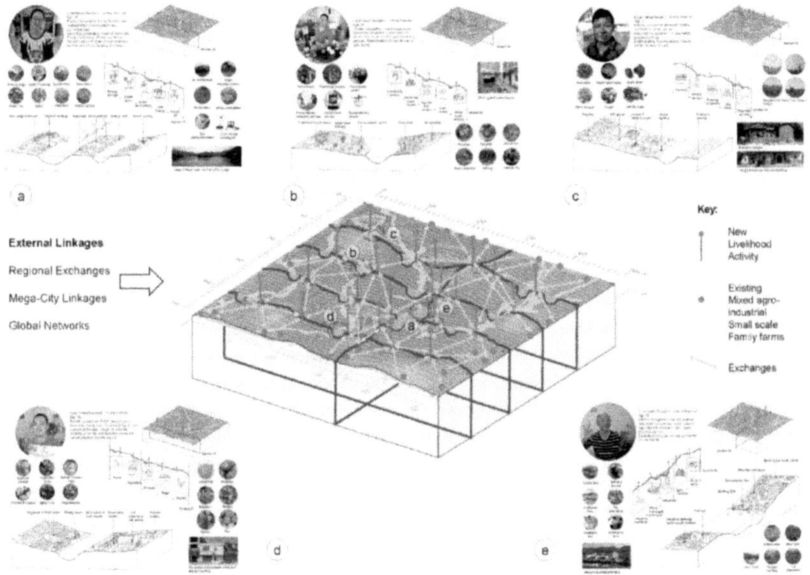

Fig. 3.96f: The multiplication, networking and exchanges of new livelihood activities will overtime grow emancipated local ecologies. Source: By the author.

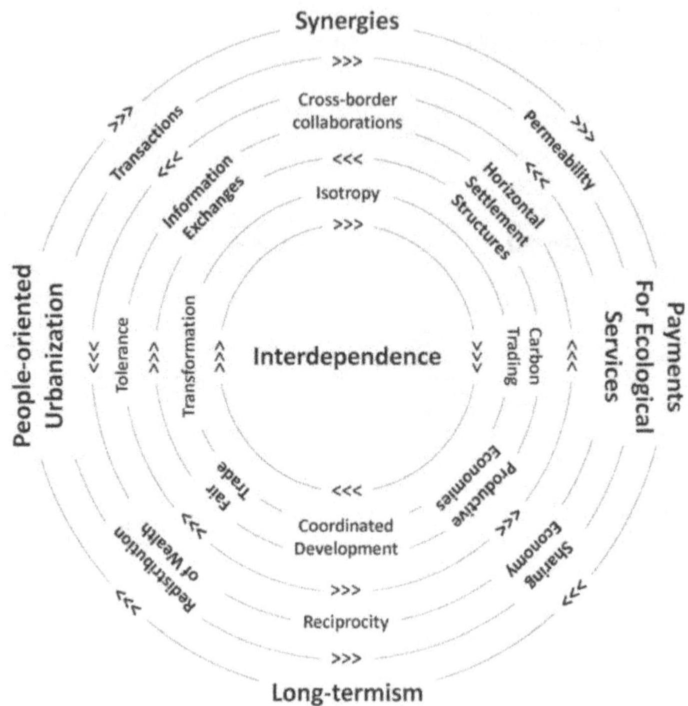

Fig. 3.96g: While territories must sustain systemic independence to a certain degree, interrelations with different urban agglomerations are the feature of an internationalized economy. For the maintenance of the long-term stability for both megacity agglomerations and dispersed mountain communities Interdependence between territories is an unavoidable necessity. The strategic preconditions for the coexistence of different types of urbanities in a globalized world are shown. Diagram: By the author.

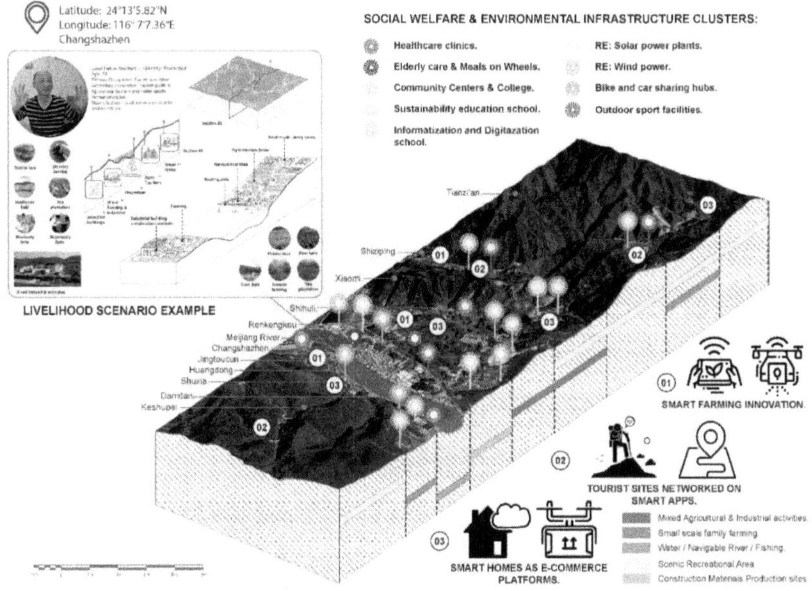

Fig. 3.97a: Scenario construction Changsha town: The case of Mr. F's livelihood strategy is presented in the context of Changshazhen. Mr. F. is a guide for visitors to agricultural-tourism and mountain scenic sites. Additionally, he works as a driver and is a drone operator for a delivery service. In the case of Mr. F's agri-tourism farms smart technologies have been embedded as shown in the section to support a future-oriented transformation of the settlement. Source: Google Earth & by the author.

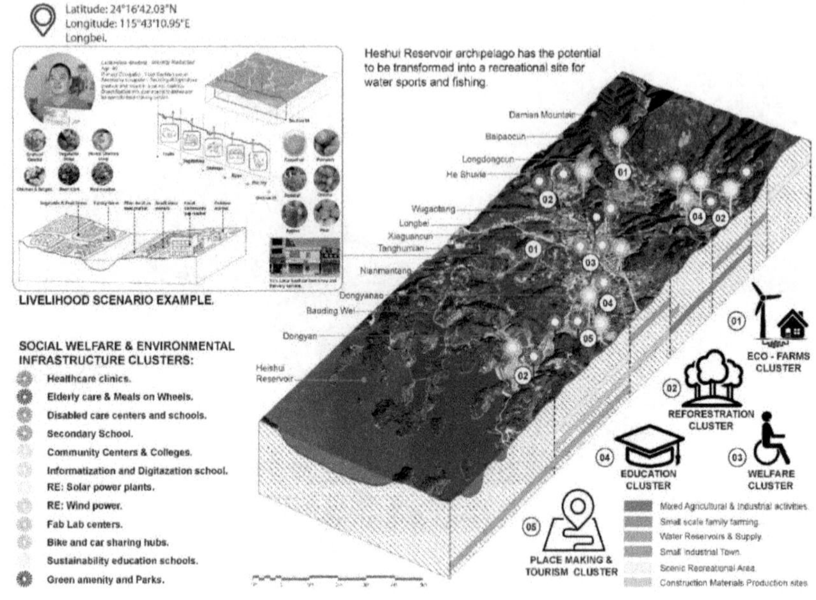

Fig. 3.97b: Scenario construction Longbei town: The case of Mr. L's livelihood strategy is taken as an example in the context of Longbei. Mr. L. specializes in Hakka cooking and distribution of farming produce. As shown in the section axonometric, his town has undergone a process of transformation by embedding several reforestation projects to restore a fragmented habitat, eco-farm clusters for sustainable farming, promotion of Heshui reservoir as a tourism destinations, the provision of schools and better welfare services for the elderly, due to its aging population. Source: Google Earth & by the author.

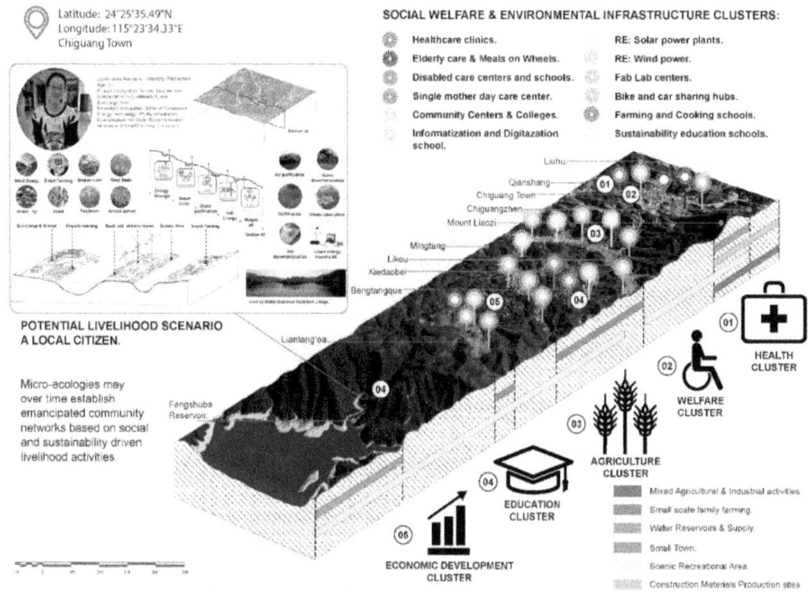

Fig. 3.97c: Scenario construction Chiguang town: The case of Mrs. L's livelihood strategy is shown an example in Chiguangzhen. Mrs. L. is a teacher by training and is interested in sustainability education. Besides teaching in the local school, she also offers advice to farmers on renewable sources of energy and water decontamination. The development strategy for Chiguangzhen is based on providing better education facilities, teaching sustainability practice for farming and energy generation in primary and secondary schools. The community also set up a clinic and nursing home to provide access to basic medical care for all members in this settlement. Source: Google Earth & by the author.

3.6. Street Transformations

To illustrate in more detail possible sustainable transformations of the settlement spaces studied, street level interventions and regeneration measures are presented in the following section. Examples of design interventions:

Chiguang town - Longbei town - Changsha town.

Street Transformations: Chiguang town.

Revitalization of Main Street as a catalyst for sustainable development of the settlement.

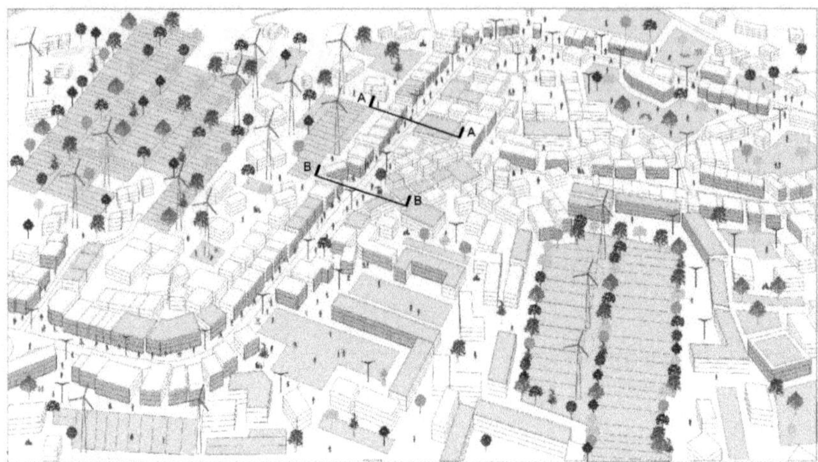

Fig. 3.98a: The revitalization of Chiguang town's main street proposes the overhauling of shop fronts, greening of roofs and installation of street lighting for improving safety at night-time and to enable the use of outdoor public spaces during evening hours for business activities and recreation. Additionally, the planting of trees, installation of vertical vegetation systems to building facades, the creation of new small parks in the back streets and underutilized plots is proposed. These interventions are intended to increase the vegetation coverage in the town. Gradual adaptation to re-newable energy generation is planned for by the incremental installation of wind turbines for the community. Farming plots are consolidated into larger fields to increase the efficiency of farming activity and secure independent food production for the town. Image: By the author.

Street Transformations: Chiguang town - Section Example A.

Key Interventions: Upgrading of local shops and creation outdoor areas for people to do business activities.

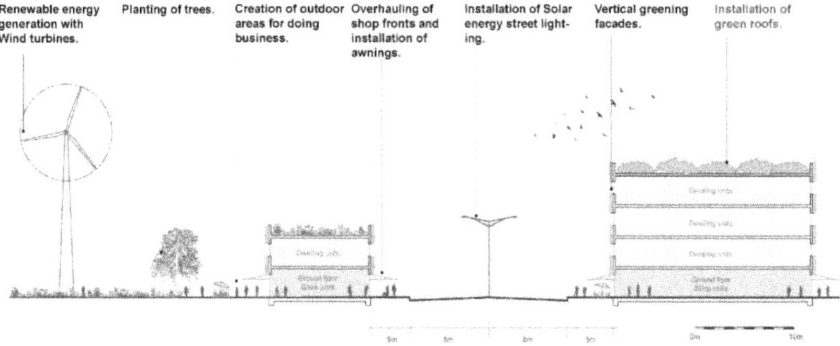

Fig. 3.98b: The section shows the overhauled ground floor shop fronts with new awnings for sun shading. New Street lighting for safety at night time is suggested. The paving zones are widened to allow for people to do business outdoors. The building facades have vertical greening and green roofs are proposed to support biodiversity in the town. Image: By the author.

Street Transformations: Chiguang town - Section Example B.

Key Interventions: Setting-up of a new Community Walk-in-clinic & a new outdoor market place.

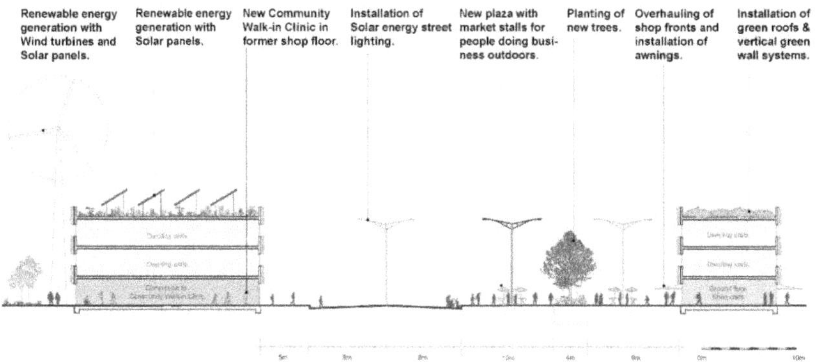

Fig. 3.98c: This example of transformation illustrates the conversion of an abandoned ground floor shop into a Community Walk-in clinic. The idea is to provide basic medical care to members of the community. To attract the local inhabitants and to promote public health, the clinic is located on the main road of this settlement facing a community a market place for easy access. The newly created market place has tree planting to create a pleasant gathering place. On the plaza, market stalls can be installed for selling local farming produce and food which people can consume at outdoor canteens. The roof areas of the buildings are utilized for the installation of solar panels and green roofs systems. Image: By the author.

Street Transformations: Longbei town.

The creation of Farming Parks engulfing the town for the development of Agricultural Tourism.

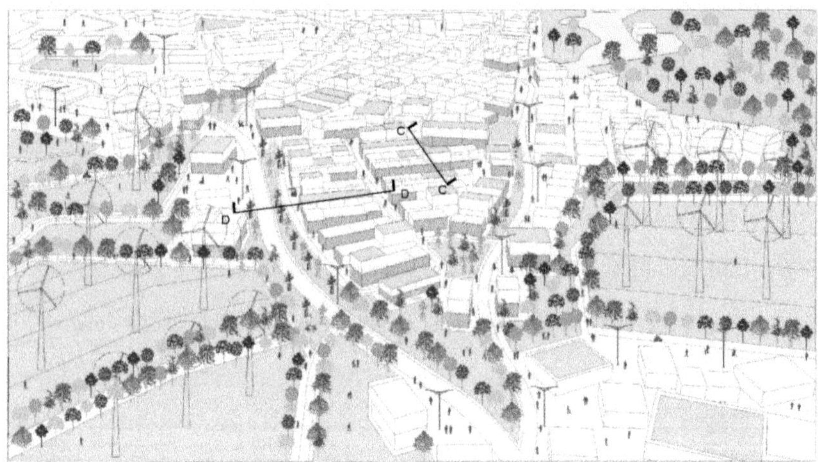

Fig. 3.98d: For the transformation of Longbei town, new farming parks for agricultural tourism are shown in this example. The aim is to remediate the environmental deterioration and at the same time to create a tourism destination. The new agricultural parks engulfing the town are intended to generate revenue from tourist, and to secure independent food production, and the selling of the agricultural produce. Fruit tree parks are popular among families from larger cities whether people have limited opportunities to experience and agricultural activities and consumption of organic food on farms. Additionally, it is envisaged to upgrade the retail lanes in Longbei town and to create new linkages in-between the narrow alley ways by combining several shop units to create larger indoor passages. Image: By the author.

Street Transformations: Longbei town - Section Example C.

Key Interventions: Narrow and dark alleyways improved with Lighting and Vegetation concepts.

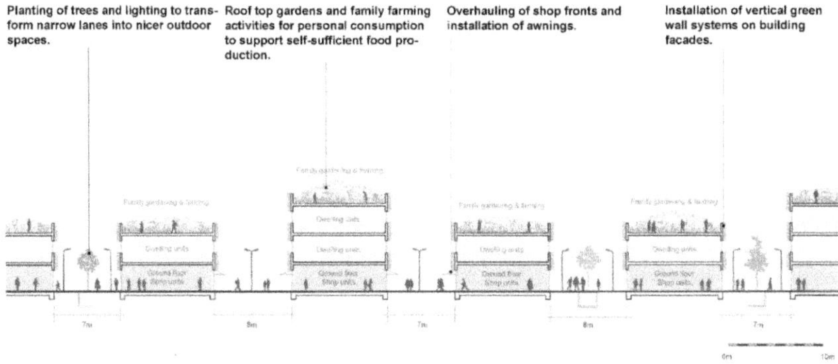

Fig. 3.98e: The narrow alley ways are transformed into upgraded shopping streets by planting trees, installing street lighting and awnings at shop fronts. The smaller roof tops are converted into gardens for inhabitants living on the upper levels of the small live and work buildings. It is suggested that some farming activities for personal consumption could be carried out on roof the gardens in addition to running a shop on the ground floor. Image: By the author.

Street Transformations: Longbei town - Section Example D.

Key Interventions: New Sport & Health centre, an Indoor Market and Food hall.

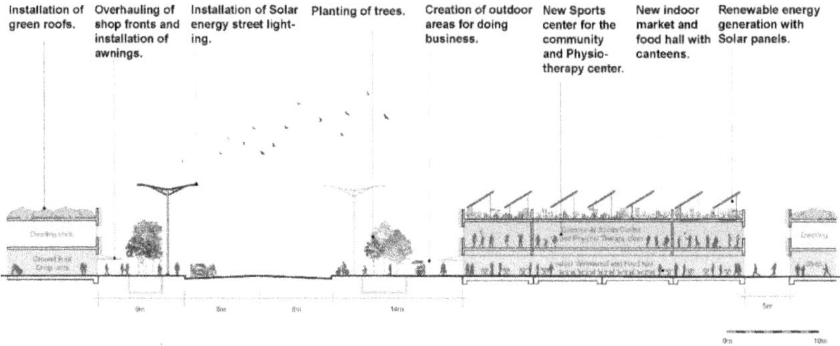

Fig. 3.98f: In this example two interventions for the community are proposed, which is the creation of a new indoor wet market, a public sports facility and a physiotherapy center for local inhabitants. Abandoned small buildings are combined for the reuse as larger indoor spaces for the sports facility. Here people can gather for team sports and exercise classes. Additionally, physiotherapists have training and consultation rooms. The upper floors facility is intended to improve the well-being and health of the inhabitants of this town. On the ground floor, the joined-together shops create new linkages between existing streets and an indoor market place with a food hall. Image: By the author.

Street Transformations: Changsha town.

Green Open Parks as In-between Linkages between River bank, Town and Farming fields.

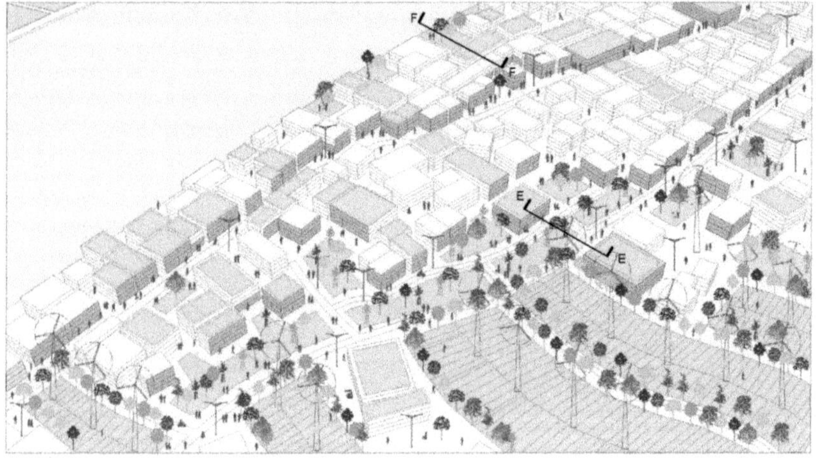

Fig. 3.98g: The mixed-use industrial and farming community of Changsha town is organized in streets running parallel to the Meijiang River. However, the settlement has its main shop front streets facing away from the river front. Additionally, the build-up area has limited linkages to the farm land behind the buildings of this town. The design proposal creates green parks, green roof areas and the planting of new trees in-between the existing buildings. This is to connect the cultivated farmlands to the revitalized shopping streets with intermediate open green park areas as much as possible. In other words, the idea is to create a better mixed transition between the built-up zone and the farming plots by introducing new parks in-between. The shops of the two-three story buildings are redecorated and renovated to be attractive to tourist and the local

community. Further, new smaller parks along the river bank are created to draw people and recreational activities to the river front. Image: By the author.

Street Transformations: Changsha town - Section Example E.

Key Interventions: Creation of a "Green Lung" in between buildings on an empty plot of land.

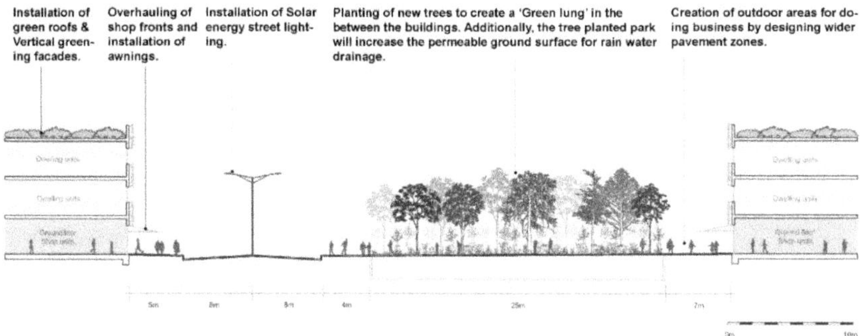

Fig. 3.98h: This street transformation section shows the creation of a "Green Lung" in-between the buildings on an undeveloped plot of land. The newly planted trees contribute the absorption of pollutants in the air and the cooling of temperature in the summer by providing shading. The shops are refurbished and a 7m paving zone between building front and the new park allows for restaurants and coffee shops to offer outdoor seating areas looking out at the new tree planted park. Vertical green walls are suggested for the existing buildings and the roof areas are converted into gardens. Image: By the author.

Street Transformations: Changsha town - Section Example F.

Key Interventions: New Kindergarten & park facing the river bank, a Mother & Child care centre and a Community Library with internet access.

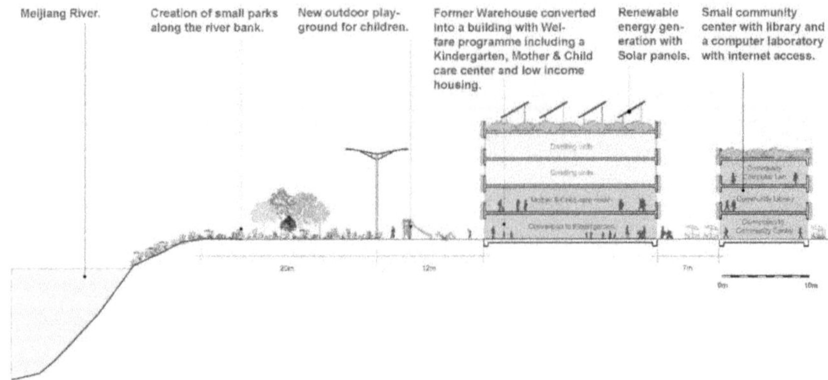

Fig. 3.98i: The river bank site has been transformed into a park facing a former warehouse retrofitted into a mixed use building with a kindergarten on the ground floor, a Mother & Child care center for families in need on the first floor. The upper floors have been converted into low income housing flats. The building is partly serviced by renewable energy from new solar panels installed on the roof. Opportunities for creating roof top green areas are maximized. A 7m wide street provides access from the kindergarten to a three-story building dedicated to a small community center with a library and computer laboratory. The community center gives people free access to the internet and computer literacy training courses. Image: By the author.

Part Four

Critical Evaluation of the transect as a planning method.

4.0. Critical Evaluation of the rural-urban transect as a method for the analysis and design of territories.

Introduction:

The Transect approach is fundamentally important in the design and analysis of territories. This critical evaluation of this planning method considers the Transect as a design tool. While the idea of the transect design tool has been common in Europe and North American planning discipline, and recently in China, the literature on the rural-urban territory continuum has been growing steadily since the 1980s.Due to the internationalisation of practices in urban planning and also exchange programmes between Chinese and foreign universities, the transect is growing as a tool for design and analysis of territories. For an effective use of this tool in the Asian context, awareness of the critique of the transect method, in particular form the North American New Urbanist Movement (NUM) which resulted in standardized suburban settlement planning is discussed. A review on the transect is best informed by an exploration of the different types of regions, i.e. rural, corridor, and metropolitan regions. Such essential knowledge is important for an understanding of the six different transect zones as elaborated by the Congress for New Urbanism in 1993. The zones are natural, rural, sub-urban, general urban, urban centre, urban core zone and the two most advanced, which are Civic zone and special districts.

The transect approach is a significant design tool compared to the plan-based zoning master planning option. The method is a "geographical cross-section of the region", showing the sequence of the different types of environments. As a method, the Transect is a design tool and a sampling, analysis and inventory technique. By comparison, the approach helps in estimating and evaluating the ecologies and characteristics of the natural resources found in a series of sections of a territory. Especially, the transect design tool can be applied to reconcile &reconstruct city morphology by embedding design ideas into the process of designing and analysing territories. The design tool can also analyse and assist in addressing the ongoing environmental pollution. For example, the transect studies enable scenario constructions which problematize environmental remediation, modern agricultural techniques and renewable energy production that foster sustainable community development. The transect method is also used as a formal framework to embed new and diversified livelihood activities in community planning, aimed at the alleviation of poverty in remote rural areas.

The origin of this geographic concept of the transect as a design tool can be traced to the Geddes Valley Section (see Fig. 4.07).The idea of the Valley Section shows the relationship between occupations of the section's residents and their corresponding settlements, which is in line with their landscape (Geddes,1915; Melemis and Tixier, 2014). The concept of rural to urban continuity evolved in Europe and according to Geddes (1915), the Valley Section is found in almost every location, i.e. in Western Europe and Britain, making up essential and characteristic geographic units. Providing a link, the in Asian context, Canadian geographer Terry McGee (2010) who introduced his Desakota concept (see Fig. 4.02) to explain the emergence of the hybrid rural – urban spaces in regions surrounding cities (Shane, 2019).This part focuses on the rural-urban transect method and approach of understanding, analyzing, and intervening in modern-day cities. The theory re-emerged with the New Urbanists in the early twenty-first century (Holmes, 2017). The rural-urban transect as a design tool has a long history in the planning discipline in Europe and North America. In particular, Behzadfar, Alalhesabi, and Amirhodaei (2017) note that the theory has become the basis for planning and designing in many urban areas and neighbourhoods in the urbanization system in America and Europe. Moreover, the application of this theory or method has extended to include various cities in Southeast Asia. In the Chinese context, the approach is still relatively new. However, the design tool is gaining increased application due to the internationalisation of planning practices and academic exchanges between China and foreign universities. This discussion starts by exploring the rural-urban transect as a planning tool that has a rich history. Then, an overview of the concepts in regional design by Neuman (2000 and 2018) is presented in this section. A critical description of the concepts of transect as a design tool follows.

History of the Rural-Urban Transect as a Planning Tool

New Urbanist planners and architects have popularised the transect over the past 15-20 years. However, the Transect as a design and planning tool has a rich history, particularly in the European and North American planning discipline. The history of the Transect spans over two centuries, first emerging when Alexander von Humboldt, a Prussian geographer, used the cross-sectional technique to diagram the earth's crust as an 'Physical Atlas' in 1841 (see Fig. 4.01). In the article The Urban Transect: One of the Planning Profession's Most Powerful Tools, Deal (2017) notes that the transect system has a storied history since it has been applied as a tool to study how nature and built environment intersect. Deal (2017) further explains that the tool has for a long time been re-imagined as an apparatus for regulation, thus providing a systematic methodology in the process of evaluating local policy choices. As Melemis and Tixier (2014) state, the idea of the transect dates back to the eighteenth century figure, the surveyor (the arpenteur), who is a scientist trained to become a surveyor-cartographer.

Fig. 4.01: Diagram of a cross-section of the earth's crust. Alexander von Humboldt was a German 19th Century Enlightenment thinker, explorer, geographer and polymath, who used large-scale geographic sections to explain and map geological information and ecological knowledge. Source: Alexander von Humboldt, 1841.

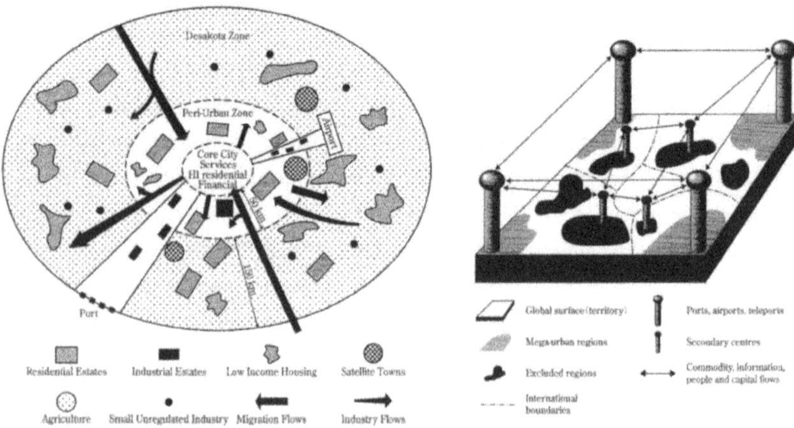

Fig. 4.02: In the East Asian context, Desakota has been coined by Terry McGee, which describes an extended metropolitan region of interlinked networks of rural-urban transactional spaces. The illustration to the right by Terry McGee shows a transactional space in a globalized economy. The diagram to the left shows a spatial arrangement of hybrid and mixed activities in a mega urban region as described by McGee, 2008.

A prominent scholar who utilised the Transect was the Scottish planner Patrick Geddes who used the concept to portray how cultures emerged as a result of existing geographical context (Geddes, 1915). In particular, the Geddes Valley Section illustrates the relationship between the environmental resources and their extractions by residents pursuing livelihood activities. Motivated by a synthesis of ecology and landscape design practice, Ian McHarg, a landscape architect, applied the Transect as a tool to analyse and design territories in his book Design with Nature (see Fig. 4.04). The Transect arose from the need to break down the city into fundamental urban typologies and forms (Deal, 2017). Some of the early professional urban designers and planners were from Europe and North America, hence evidence of a long history of the concept in these regions. As mentioned earlier, the concept of the Transect is relatively new in Asia, particularly in China. The literature detailing the continuum in rural-urban territories is an increasingly growing research field. Partly, this is due to the objectives of the CPC's National New Type Urbanization Plan 2014 – 2020 in China (Xianping, 2014), which emphasises a sustainable approach to urban and rural development. Moreover, due to the internationalisation of practices in urban planning, and also exchange programmes between Chinese and foreign universities, the transect as a design tool is receiving increased attention and application in design practice.

Traditional zoning strategies emphasised on discerning urban patterns. However, the differentiation of settlement types and regions increased, and there was inadequate rigidity and separation in land use (Endlicher et al., 2007). This prompted many urban planning professionals to gain more interest in alternative models that fully capture the fundamental patterns and design the principles that cities required to thrive. The re-discovered model, i.e. the Transect, signified a "back to basics" approach in such a way that city zoning regulations could lead back to the earliest plans of zoning. For instance, Emily Talen, a professor at Arizona State University, after fully studying the earliest zoning codes in the country, noted that the guiding assumptions and framework that characterised the early plans are similar to the modern codes produced by the contemporary transect framework (Deal, 2017). Some planners and designers called for the development of a simpler as well as transparent process for land use coding. The first transect, i.e. the rural-urban transect, was designed for use by local government officials (Deal, 2017). In the 1980's, the concept of transect as land use, evolved as part of the NUM in North America, also known as the SmartCode (see Fig. 4.06), will be discussed in a latter section. The following section describes regional design. The description is essential as it helps in attaining an overview of the spaces and scales for designers working on regional projects and how it assists in portraying the relevance of the transect approach as a method of working in territories.

The Geddes Triad of PLACE - WORK - FOLK.

The diagrammatic representation to the left by Patrick Geddes illustrates the correlations of Place, Work and Folk. It proposes a structured study of human interactions and their environment. This model has been adapted in the diagram to the right for this study to shift the attention to Ecology, Livelihood and Community. This is to show the interrelatedness of the three categories applicable to large, medium and smaller settlements. The relationship of human livelihood activities relative to territories and the ecology can be analyzed and represented in the Geddes Valley Section.

PLACE	PLACE WORK	PLACE FOLK
WORK PLACE	WORK	WORK FOLK
FOLK PLACE	FOLK WORK	FOLK

ECOLOGY	ECOLOGY LIVELIHOOD	ECOLOGY COMMUNITY
LIVELIHOOD ECOLOGY	LIVELIHOOD	LIVELIHOOD COMMUNITY
COMMUNITY ECOLOGY	COMMUNITY LIVELIHOOD	COMMUNITY

Fig. 4.03: Geddes triad of WORK, PLACE and FOLK from Cities in Evolution: an introduction to the town planning movement and the study of civics, Williams & Norgate, 1915. The illustration to the right shows a triad of Ecology, Livelihood and Community is an assimilation for this study by the author.

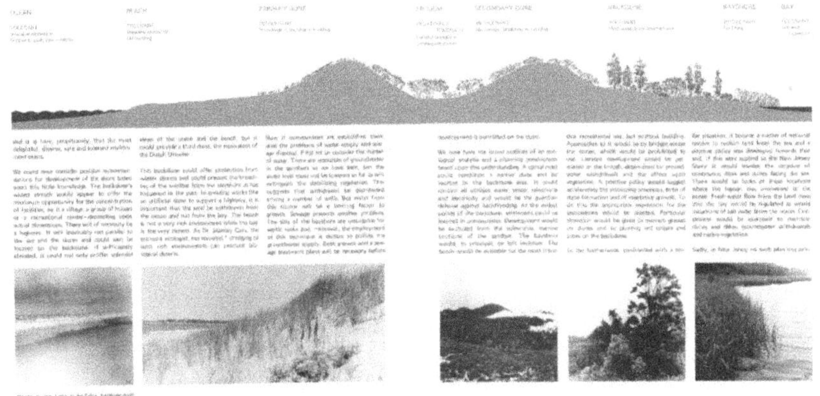

Fig. 4.04: A transect by Ian McHarg from his book Design with Nature published in1969. As a landscape architect McHarg advocated for ecological planning as an intrinsic part of regional planning. The transect is used to coordinate ecological planning and land-use planning n multiple overlapping ecological systems: Ian McHarg, 1969.

An Overview of the concept of Regional Design

Before embarking on the idea of regional design, it is necessary to define the term region. Defining regions to suit the purpose of urbanism requires consideration of the object of planning. Sustainable communities are the key objects of planning, but it is also essential to look at the settlement structures, as communities are found within larger-scale spatial configurations. The specific topography and climate of a region, for instance, significantly influence the evolution of settlements and morphology (Neuman, 2000). From the perspective of land use planning, a region is an inter-connected territory that its residents relate to through their economic transactions and activities. A region can also be described as an area where individuals live and execute most activities that are recurring, or in other words, pursue their daily-lives.

According to Neuman (2000), the regional design is an approach to embrace growth through the provision of a physical framework that helps in guiding or determining the best use of land in terms of shared benefits, the scale, and resources as well as understanding how communities interrelate within a particular physical region. Regional design as a strategic design discipline is unique or different from that of urban and regional planning, but both disciplines have an impact on the morphology and transactional networks (Mc Gee, 2008) in a territory. Neuman and Zonneveld (2018) assert that regional design guides human settlements in a particular area through the definition of specialized functions, size, location, and how settlements interrelate. In addition, regional design directs the flow of human endeavours in relation to settlements through the existing linkages in infrastructure. The ultimate goal is to address a better integration of infrastructure networks and settlements with eco-system services; hence attaining sustainability (Neuman and Zonneveld, 2018).

The material design of a territory (Viganò, Fabian and Secchi, 2016), both in theory and practice, is increasingly re-emerging as design problems in contemporary urbanism. In another article, Neuman and Zonneveld (2018) give two main reasons for the resurgence of regional design. From a structuralism point of view, it can be argued that regional design has been a longstanding basis for spatial planning, and it has re-emerged as it is a tool for spatial strategy, and it also serves in spatial management.

Neuman (2000) also note that the rebirth is as a result of neo-traditional community planning and also modern-day urbanism, two concepts that started as enterprises that initially utilised small scales, i.e., neighbourhoods and small communities. However, with the evolving knowledge in urban planning and design, the scale has grown to incorporate large cities that are part of mega-city regions. This is in addition to the dispersed rural-urban areas that overtime evolved into towns and low population cities. The essence of regional design can best be understood by considering the emergence of new settlement structures with a regional perspective and also acknowledging the global external factors or forces that drive urban growth. One such indicator of the forces and context, for instance, is the rise and growth of metropolitan economies as influential nodes in a service- based global economy.

The need to adapt to the new realities in the evolving world emphasises the essence of regional design. The new realities include high-tech digital innovation in telecommunications, the internet, and information systems that enable exchanges of capital and resources with metropolitan economies, which ultimately network a particular region. In addition, there is increased mobility, i.e., national and international migration, tourism, international business, and also intra-regional mobility, which include increased emergence of plural modes of mobility, such as autonomous driverless cars and drone transportation. New conditions such as an increase in urban population, climate change, globalisation and their associated impacts, 'smartcity' concepts and new infrastructure technologies drive urbanisation. With the new conditions of urbanisation, the need for forming as well as applying spatial strategy at supra-metropolitan levels or scales cannot be overlooked (Neuman and Zonneveld 2018). The new forces that drive urbanisation in the contemporary and future world represent a new scale of a dispersed urbanity, especially in the city-region. Neuman and Zonneveld (2018) assert that the forces of urbanisation are associated with significant impacts that need to be addressed or dealt with by urbanism at a large scale that matches a comprehensive urban- to-rural space. Regional design represents the strategic solutions to the macro-scale design problems in urbanism.

Regional design is pertinent in shaping the physical form of regions considering it embraces a regional viewpoint that guides how human settlements, particularly in diffused cities (Indovina and Matassoni, 1990), are arranged. With the new realities, regional design connects the communities through communication, transport, and such other ways that cause urban- rural networks to emerge. Similarly, regional design helps in keeping the territory of the communities sparsely populated, by a better coordinated use of resources and by avoiding an over-emphasis of high-density urban developments, for instance. Neuman (2000) proposes three objects of regional design, namely communities, the connections among them, and their environment. Another key aspect when assessing regional design pertains understanding the contemporary design problems in urbanism.

According to Neuman (2000), some of the sources of current regional design thinking are the existing critique of urban sprawl, the need to retrofit suburban settlements (Dunham-Jones and Williamson, 2008), the planning of resilient settlements in an evolving world, the idea of the countryside as a city, and the notion of mountain settlements as an alternative urbanity. In the Chinese context, the concept of a 'harmonious' development of communities is attracting research attention (Edmonds, 2012). The key concerns are the social problems and the environmental degradation found both in the Chinese countryside and cities. The acute income inequality between urban areas and the poverty-stricken mountain regions in China is an urgent issue to be addressed. The next section discusses different types of regions, according to Neuman (2000), and acknowledges how regional design and design of territories interact. There are different types of regional settlement patterns. Neuman (2000) informs that basic regional settlement patterns are the rural hinterland, corridors, metropolitan; but in other literature, authors also refer to mega-city regions, urban fringe areas, mega-city corridors, 'rurban' spaces in-between and the mountain territories. Regardless of the evolving types of regional settlement patterns, a region has the same definition i.e. a network of communities that can be concentrically-organized, polycentric, or dispersed, and which are linked by post-fordist, modern communication and transportation systems and whose surrounding is land that is less densely settled. A significant concept pertains command centres, which refer to urban infrastructure of places serving a network of global cities (Sassen, 2004). Sassen (2004) further expounds on the significance of command centres. As the author explains, global cities such as Tokyo, New York and London are linked to each other and to financial command centres like Paris, Los Angeles and Hong Kong than they are connected to their own provinces and states. In this way, the shift in the sense of what a globally integrated region compared to a locally isolated region entails is evident. Metropolitan regions are urbanized areas that tend to be densely populated and their infrastructure i.e. roads, buildings and spaces are interconnected by an 'urban fabric'. Such regions are increasingly diversified compared to the rural and corridor regions (Neuman and Zonneveld, 2018). Corridor regions, on the other side, are areas that surround linear routes of transportation or a river for example (Neuman, 2000). A common aspect of the corridor region is the corridor for transportation or watercourse that acts as the spine or focal axis of the area. In terms of scale, morphology and function, corridor regions vary. For instance, the Greater Bay Area project in China partly constitutes of three key infrastructural extensions which are the Guangzhou-Shenzhen-Hong Kong Express Rail Link, Hong Kong-Zhuhai-Macau Bridge and the Northern Lian Tang/Heung Yuen Wai Checkpoint as a linkage between Hong Kong and Shenzhen. Similarly, Neuman notes that the lands along corridor areas represent the diverse patterns of settlement of the areas they comprise. As an example, the Beijing – Tiangjing– Hebei integration project in China, aims to rebalance the problem of industrial overcapacity building and duplication of urban functions in mega-city clusters by introducing infrastructural and economic corridors in the region to better share resources.

While acknowledging the binary simplification of 'city' and 'countryside', a third basic settlement patterns is the 'rural' region. Neuman (2000) states that these regions typically consist of hamlets in the rural areas, villages and small towns that have open lands. Compared to the metropolitan regions, rural regions are characterised by farms, forests, water reservoirs, and other open lands such as marshes that comprise the environments around rural settled places. The rural settlements may lack advanced transport, communication or utility lines as the settlement is sparse in the surrounding environments but denser in rural communities and small towns. Increasingly, a mixture of agricultural and industrial activities relying on state-of-the-art know-how brought from urban areas characterise rural regions. Other types of rural regions are ecological reservoirs or uncultivated land. In discussing types of regions, hybrid regions should also be considered. In China, urban villages developed within cities as a result of rapidly growing urban expansion engulfing rural communities. Hybrid regions begin to form when different region types, i.e. metropolitan and rural, intermesh or merge. An example of a European corridor megalopolis is found along Germany's Rhine River and extends to Dortmund from Stuttgart (Neuman, 2000). The megalopolis region depicts a large-scale metropolis region. Knowledge on the different types of regions is essential to understanding how the various regional types impact on interrelationships of communities in a particular region. The knowledge also helps in understanding the linkages that bridge the regions as well as the intermediate territories that separate or buffer them. These are key aspects in regional design. Regional design and planning is dependent on the scale and spaces of regions. As Neuman and Zonneveld (2018) note, cross-border, transnational planning at the continental and supra-national scales started in Europe. The planning can be dated back to the origin of spatial planning in other developed nations such as France, Germany, and Italy. In the past twenty years, planning focused on new scales, i.e. trans-national, cross-border and even European which stimulated approaches to regional design founded on innovation and creativity necessary to intervene in extended territories (Neuman and Zonneveld, 2018). In France, 'L'Organisation de l'espace: Eléments de géographie volontaire' was published by Jean Labasse in 1966. The concept of 'la géographie volontaire' was pursued as comprehensive approach for the management of territories by architects, geographers and urbanists motivated by the need for regional reconstruction projects after WWII (Cupers, 2016). In Italy, architect Vittorio Gregotti called for expanding the design scope of architecture to include territories in his book Il territorio dell'architettura from 1966 (see Fig. 4.05).

Neuman and Zonneveld (2018) further posit that regional design should be complemented by innovations of the existing local institutions and by community-level participation. With this endeavour, there are increased chances of filling gaps in modern- day spatial planning through the development of regional policies, development programs, and integrative processes for citizen that are effective. The desired effect is establishment as well as the attracting of financial resources and development processes that will favour investment in infrastructure. Ultimately, it may help to reduce regional inequalities, increased protection of farm land and ecological resources. At the regional scale, innovative forms of legal and policy instruments could be utilized to attain sustainable development. Such instruments could include Payments for Ecological Services (Zhang, Bennett, Kannan and Jin 2010), basic farm land preservation for food independence, and land use reforms for agricultural land. For example, additional development policies and schemes include the controversial land coupon programs (Han and Lin, 2019) implemented in Chongqing, China, emissions trading schemes tested in Shenzhen (Ye, Jiang, Miao, Li and Peng, 2016), access to micro-credits loans for rural communities, infrastructure finance and value capture schemes at the regional scale that serve to spread the benefits resource utilization and environmental trade-offs caused by development and re-development (Neuman and Zonneveld 2018). Regional design offers a toolbox that designers, planners, and policymakers could use to overcome various objections to local and national planning. With the above information on the spaces and scales in regional design, it supports to highlight the significance of the Transect as a method for designers working on regional projects. The next section discusses the transect approach as a method for the analysis and design interventions in territories.

4.1. Epistemological Traditions: Géographie Volontaire, Urban – Rural territories, and the Garden City Movement.

The application of Design Thinking to larger scale spatial configurations. A 'scaling-up' of research by design knowledge production. (K.Cupers, 2016).

Fig. 4.05: L'organisation de l'espace by Jean Labasse, and It territorio dell' architettura by Vittorio Gregotti advocate the expansion of the scope of the architectural discipline to design in larger scale territories. The origins of this hypothesis can be traced back to post WWII European national reconstruction efforts of economies. Given the current Global Challenges including climate change, food insecurity, poverty and environmental crisis research projects in larger scale territories are recognized as urgency for designers.

Merits and Contradictions of the Transect as a Method for the Analysis and Design of Territories

The section considers transect as a method or approach of analysis, with a focus on design and analysis of territories. The transect method is not only an affordable but also a fast data collection technique. Before discussing transect as a method, it is pertinent to elaborate on the different definitions of the term transect. Both the benefits and critique of the transect method promoted by the NUM will be discussed. A transect can also be described as a way of locating and acknowledging different types of human settlements in a network of human and natural habitats (Bohl and Plater-Zyberk 2006). A transect, according to the SmartCode Manual by Duany Plater-Zyberk and Company, refers to a geographical cross-section of an area or region, and it is used to show a sequence and continuity of environments (Duany, Sorlien, and Wright 2005). The SmartCode Manual further explains that transects were originally applied to analyse natural ecologies, and they showed different features via various zones such as wetlands, shores, plains as well as uplands. For human settings, the cross-section helps in identifying a number of habitats which vary in regard to the intensity of land use functions and elevation difference relative to the urban region, and this continuum ranges from rural-urban. Duany, Sorlien, and Wright (2005) further inform that in Transect planning, the different settings or environments are the foundation for organising the various parts of urbanisation i.e. building land use, lot, street, and other physical features of the human settlement.

Experts in urban transect approach hold two major perspectives towards the rural-urban concept. These are in the form of functional and also theoretical frameworks. According to Bohl and Plater-Zyberk (2006), the transect can be perceived as a theoretical framework and as a functional model or framework. Urban transect as a theory or approach pertains the view on the morphology of the urban fabric, porosity of built up areas and urban green space provision. On the other hand, analysis of rural- urban regions using an ecology and functional framework seeks to understand the possible correlations; hence, the focus of this perspective is the discovery of ecological continuities, detection of fragmented networks, potential environmental risks and sustainable development opportunities of territories.

Analysing the rural-urban transect approach as the geographical cross-section considers the urban-rural gradient as a form of the cross-section. This urban-rural gradient is seen between the built and the open or natural areas. In other words, viewing the transect approach as a geographical cross-section means that the rural-urban differentiation lies in the variations between the built regions, i.e. 'urban' areas and the 'rural' regions that are open and natural. Beyond the binary perspective, this point of view, offers a gradual transition between the two settlements (Neuman 2000). According to Bohl and Plater-Zyberk (2006), the gradient and fluctuations in rural-urban physical features or specifications are shown along with the ongoing evolution or urbanisation. Such specifications include population, abundance or density of the urban areas, form of the impenetrable surface cover, structure of the population, and also attitudes and lifestyles choices.

In particular, the Geddes Valley transect model enables to represent a differentiation between rural-urban areas, and this helps in connecting the region with the topography and environmental capital, local ethnic characteristics and communities in different regions. The transect method considers the existing order of the natural environment in the built environments and further extends it into the core of the metropolitan area (Behzadfar, Alalhesabi, & Amirhodaei, 2017). In this instance, the rural- urban transect model analyses the existing biodiversity in the built up fabric; hence, the city remains an inseparable component of the transect spectrum or sequence.

In parallel to the evolution of the transect by Sir Patrick Geddes (1854-1932), social reformer Ebenezer Howard (1850- 1928), who became a prominent urban planner in the UK and founder of the Garden City Movement brought environmentally and socially-conscious ideas to his projects. His garden city concept has become highly influential and continues to inspire contemporary urban planning project. His visions have been illustrated by town planning diagrams (see Fig. 4.05a-4.05c), and his book 'A Peaceful Path to Real Reform', first published in 1898. As sustainable urban planning is recognized as an urgent priority internationally, some recent projects looking at urban-rural hybrid territories, environmentally-friendly, and people-oriented urban planning projects in China are presented in Fig. 4.08a-4.08z.

The Garden City Movement - combining the benefits of town and countryside communities.

Garden City Principles:

"To-morrow: A Peaceful Path to Real Reform" - Ebenezer Howard, 1898.

- Strong vision, leadership and community engagement.
- Land value capture for the benefit of the community.
- Community ownership of land and long-term stewardship of assets.
- Mixed-tenure homes and housing types that are affordable for ordinary people.
- Beautifully and imaginatively designed homes with gardens in healthy. communities.
- A strong local jobs offer in the Garden City itself and within easy commuting distance.
- Opportunities for residents to grow their own food, including allotments.
- Generous green space, including: surrounding belt of countryside to prevent unplanned sprawl; well connected and biodiversity-rich public parks; high quality gardens; tree-lined streets; and open spaces.
- Strong cultural, recreational and shopping facilities in walkable neighborhoods Integrated and accessible transport systems.

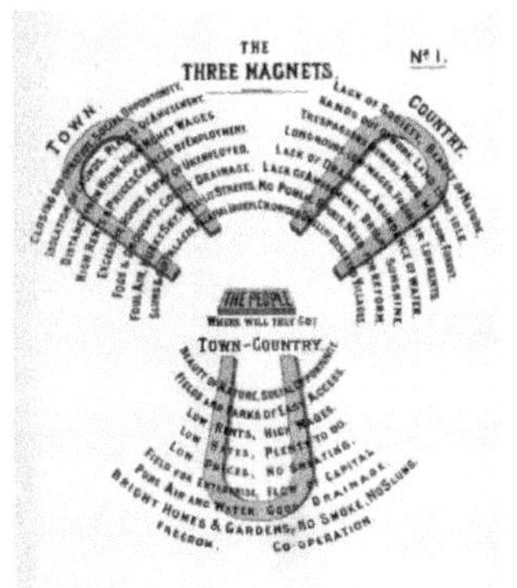

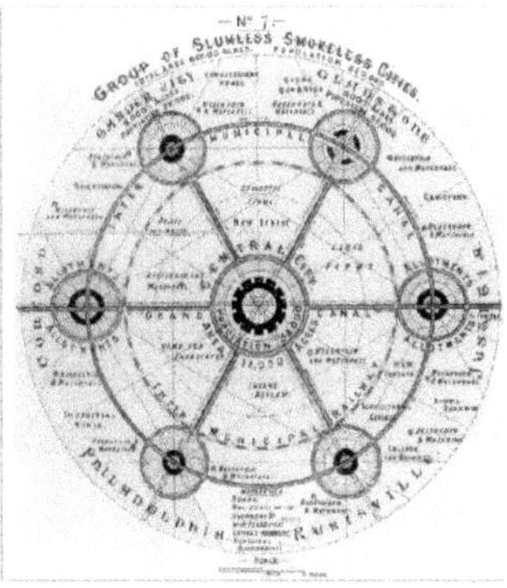

Fig. 4.05a: The Three Magnets diagram: Town, Country and Town-Country and 'Group of Slumless, Smokeless Cities' diagram. Sources:"To-morrow: A Peaceful Path to Real Reform" - Ebenezer Howard, 1898 and the Int.Garden City Institute.

The Garden City Movement - combining the benefits of town and countryside communities.

Fig. 4.05b: Welwyn Garden City in Hertfordshire by Howard, an example for better community planning. Source:"To-morrow: A Peaceful Path to Real Reform"- Ebenezer Howard, 1898.

The Garden City Movement - combining the benefits of town and countryside in communities.

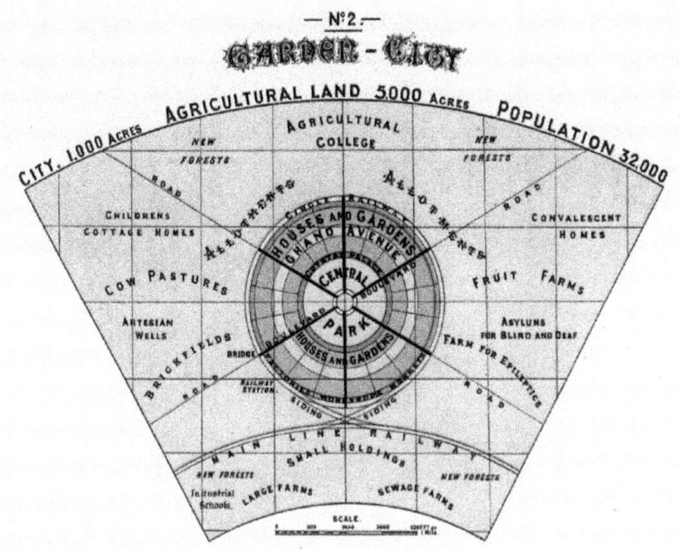

Fig. 4.05c: The diagram for a garden city shows 1000 acres of town land and 5000 acres of agricultural land. (6,000 acres or 2,400 ha). A garden city would be planned for 32,000 people. A cluster of six satellite garden cities surrounding a town center of 50000 people would form a group of garden cities see Figure 4.05a. Letchworth, the first garden city was built in 1899 by the Garden Cities Association founded by Howard. The objective was to provide better living conditions for the working class living in 'crowded, unhealthy cities'. Source:"To-morrow: A Peaceful Path to Real Reform" - Ebenezer Howard, 1898.

While critiqued as a rigid, prescriptive model, its usefulness as a planning framework, the Smartcode, which evolved in North America, can assist in the rationalization of large-scale regional problems. Design and analysis of territories can to a certain extent benefit from an understanding of this design system. The SmartCode is essentially a template document on planning and zoning of land based on the transect environmental analysis. The design system addresses different levels and scales of urban planning. The levels are from the region, natural and cultivated land, to community, to the block, and also to buildings. For the purpose of initial reference, the template provides guidance for local calibration of settlements and neighbourhoods. Duany, Sorlien, and Wright (2005) define SmartCode as a morphology-based code that utilises the principles of Smart Growth and New Urbanism (see Fig. 4.06).

Similarly, the authors note that the SmartCode is a unified developmental framework that details development at all levels or scales of design. In this particular transect approach, the scales of design range from regional planning to urban signage. Although, contradictory to this design system's idea of implementing its zoning rationale as a planning ordinance, the SmartCode claims to advocate a rural-urban transect instead of the separated-use zoning of the land (Bohl and Plater-Zyberk, 2006). In theory, Duany, Sorlien, and Wright (2005) maintain the code integrates all the necessary environmental approaches and design techniques. Its authors further hold, this system is preferable as it is more efficient and succinct compared to other conventional codes considering it (SmartCode) seeks to attain results based on patterns of urban-rural design that are well known (Duany, Sorlien, and Wright, 2005). As it is intended as a paradigmatic model, the SmartCode is promoted as a methodology, which is precise and research-based. Therefore, according to Duany, Sorlien, and Wright (2005) municipal planning departments could embrace the model, and the local administration's elected representatives should be trained in interpreting it. Moreover, considering it is also inclusive for the local citizens to participate in the design of the territories, the SmartCode may provide opportunities for integrating design collaborations and community involvement.

In transect approach, to some extent zoning codes can prove to be helpful. As Duany, Sorlien, and Wright (2005) explain, the traditional zoning approach separate residential from work and recreational places, schools and shops. While the codes entail standards of design that favour automobiles in comparison to pedestrians, and cannot resist the standardizing impacts of globalisation, new types of mobility & communication networks should not be overlooked. In this light, traditional city design thinking applied to mixed rural-urban territories, which utilises simple codes, is inadequate and associated with spatial planning challenges. The problems include banal subdivisions of housing, strip of shops, business areas, parking lots, large warehouses and local market plazas. The traditional approach to zoning has also made cycling or walking dangerous and ultimately, destroyed towns and open spaces, a 20th-century occurrence referred to as sprawl. The SmartCode claims it addresses such challenges at the point where urban planning guidelines and design intersect. Considering the SmartCode is a morphology- based code, it seeks to accomplish or encourages a given physical outcome, for instance, a particular urban fabric, new transportation networks, community, or even building typologies (Duany, Sorlien, and Wright, 2005).

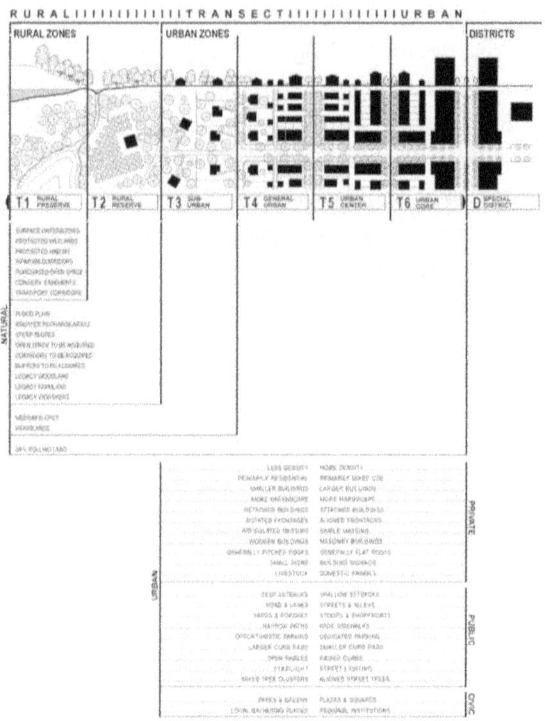

Fig. 4.06: The SmartCode developed by Andres Duany, Sandy Sorlien, William Wright uses the urban-to-rural transect as a planning tool and proposes a planning code based on structured transect zones. Although a highly idealized approach to a gradient City – to - Countryside planning tool, the systematic approach can to a certain degree assist as a framework for working in highly fragmented and hybrid urban-rural territories. Source: by Andres Duany, Sandy Sorlien, William Wright, 2008.

Yet, while the SmartCode promotes a linked network that eases traffic congestion, reliance on automobiles is necessary for commuting from NUM settlements to existing urban centers. Another, objective of the NUM is the preservation of open lands and vegetation coverage, since one of its aims is to work at the scale or level of the region and community (Duany, Sorlien, and Wright, 2005). However, the need to consume and develop open lands to set up NUM settlements must not be overlooked. Another paradox is that although the Smartcode system is conceived as a transect design tool, NUM projects tend to utilize ground-plan based methods for planning new settlements. In contrast, a transect, as earlier noted, refers to a cross-section of a particular geographical territory that portrays a sequence of environments (Behzadfar, Alalhesabi, and Amirhodaei, 2017).The transect assists designers and planners in studying the various elements that make up habitats where some animals and plants species survive in symbiotic ties to the ecosystem, including water, soil, minerals and microclimates.

There is also the need to note that human beings also choose to live in different types of places, and personal preference plays a crucial role in this differentiation. For instance, some individuals would not live in a metropolitan area, and others would never last in a rural community (Duany, Sorlien, and Wright, 2005). The New Urbanist Movement emerged at the end of the 20thcentury where architects and planners acknowledged the implications of urban expansion, i.e. the sprawl destroyed the rural environment (Holmes, 2017), which was characteristic of the American transect before WWII. The designers further enlarged the natural transect to make it incorporate urbanised environments, and this was the positive aspect of the comprehensive approach to regional design. Paradoxically, the Fordist urban sprawl and building of new suburban settlements, initially critiqued by the NUM in the 1980's, has gradually assimilated the NUM urban design and architectural styles without conceiving sustainable or autonomous communities. Nevertheless, the Smart Code design system offers helpful rationalizations of the rural-urban transect, especially when working in fragmented and disconnected territories.

Planners are most familiar with the rural-urban Transect that Andres Duany and his design company DPZ formulated. The Transect has six different Transect Zones that apply on the zoning maps. According to Deal (2017), each zone represents a single part of the large-scale urban-rural continuum. The six different habitats have varying levels and intensity of social and physical characteristics, and this offers spatial contexts that gradually transition countryside into city. The six habitats are the Transect Zones. The intention of the T-zones is to coordinate the urban design elements at all planning scales. This is from the region, community, individual plot or block and building. Despite the contradictions of the idealized, gradient transition from rural to urban areas, the NUM idea of structuring and differentiation into six typical habitats as a reference is valuable. At the same time, such typical habitats, although convenient, tend to be found as hybrid habitats in regional design, and any boundaries between habitats are often ambiguous.

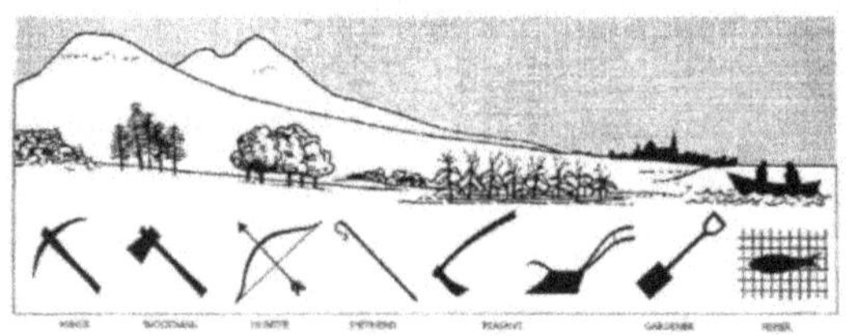

MINER - WOODMAN – HUNTER – SHEPHERD – PEASANT - FISHER

'ECOLOGIES'

ALPINE - SUB-ALPINE – ARABLE – RURAL-URBAN – COASTAL

'LIVELIHOOD STRATEGIES'

NEW MOBILITIES - INFORMATIZATION – MODERN AGRONOMY – RE-ENERGY – E-COMMERCE.

'COMMUNITIES'

MOUNTAIN DWELLERS – RURAL COMMUNITIES – 'DIFFUSED' CITY - URBANITES

Fig. 4.07: The Valley Section by Geddes in 1909, explores the merging of rural and urban. It discusses the various settlement types, extractions of resources, occupations of inhabitants in their physical environments of the region and how they interact. Geddes believes the "town arises and renews itself from the country...in character, individual and social.". Source: Patrick Geddes, 1915.

Opportunities for new livelihood strategies enhanced by new mobility systems, the informatization of regions, modern agronomy or precision farming, renewable sources of energy and e-commerce platforms have been added to the diagram. This is to suggest the embedding of new know-how and technologies in transects for the transformation of fragmented environments.

As a framework, the Transect identifies different habitats, starting from the fundamentally natural to the most developed, i.e. urban. When the continuum of the transect is subdivided, the result are gradual zoning categories. The categories entail standards that promote the diversity that equates to what is found in settlements that have evolved gradually over time (Holmes, 2017). It is pertinent to point out that the standards are interrelated; hence overlap, and this is a reflection of the successional ecotones of human and natural communities (Scott et al., 2013). In this way, the Transect as a design tool incorporates a gradient approach to zoning and also considers environmentally-friendly design interventions. As a result, environmentalists can effectively assess the design of community habitats which support the preservation of natural habitats (Holmes2017). The next section discusses the zoning categories in line with the NUM transect approach. The highly idealized Transect zones range fromT-1 to T-6, i.e. natural, rural, sub-urban, general urban, urban centre and urban core zone. Other key zones are civic zone and special districts.

T-1 Natural Zone

The zone comprises lands that can be described as having wilderness conditions. In other words, these lands are not best suitable for settlement. There is heavy vegetation cover that limits human settlement. Besides, these lands are characterized by hydrology and topography that is particularly steep and mountainous, which is challenging for cultivation (Duany, Sorlien, and Wright 2005; Bohl and Plater-Zyberk 2006). Deal (2017) share similar insights regarding the natural zone, noting that it is fundamentally an open wilderness and where humans least intrude. Therefore, environmentalists are more likely to prefer natural zone (T1) where the growth of vegetation and green cover is not impacted by human invasion.

T-2 Rural Zone

This zone consists of lands that are sparsely settled, and which are in the cultivated or open state. In T2, settled pasture land replaces nature. Examples of such zones include agricultural land, woodland, grassland, and desert that are irrigable. In such regions, it is common to find typical buildings such as agricultural buildings, farmhouses, scenic landscape areas, stand-alone cabins and Open Space areas. In line with the concept of rural design, this zone that has low-density residential locations and open space areas fall under the category of the rural regions (Neuman and Zonneveld 2018). As Deal (2017) informs, country crossroads communities and agricultural operations are the important types of human development in the section. With an understanding of the practices in rural zones, it is evident that the Transect approach contributes to finding suitable spatial location of the different features or elements that make up human residences or habitat. In particular, rural settlements tend to be distributed in the hinterland outside of cities, whereas urban development gravitates towards existing metropolitan areas to benefit from the diversified economic activity. This is in contrast to the natural ecological systems where animal and plant species coexist in habitats that fully support them (Geddes, 1915). In the Transect design tool, there is a distribution of urban development in such a way that it strengthens, instead of stressing, the integrity of every immersive setting, such as rural and open space land.

T-3 Sub-Urban Zone

This third zone consists of low-density human settlement areas. The area is close to sparsely developed zones with some mixed use (Duany, Sorlien, and Wright, 2005; Deal, 2017). Home occupations and outbuildings characterise the region. Besides, there is naturalistic planting, and at the same time, setbacks tend to be relatively deep. Moreover, blocks in the sub-urban zone may be large, and the roads are irregular for the accommodation of natural conditions.

T-4 Urban Fringe to General Urban Zone

The general urban zone comprises of mixed use, including agricultural and industrial land use. However, the overall feature or character of such neighbourhood is marked by dwellings and warehouses, i.e. residential urban fabric mixed with industrial sites (Duany, Sorlien, and Wright, 2005). The zone has a variety of building types such as side yard, single and row houses, factory buildings and workshops. In such zones, landscaping and setbacks tend to vary. Moreover, Deal (2017) explains that streets with sidewalks and curbs are the defining features of medium and small-sized blocks. The differentiation matches the idea of metropolitan regions by Neuman and Zonneveld (2018). This characteristic has some similarities with McGee's idea of 'desakota' where agricultural, industrial and urban forms of land use and settlement coexist in an urban-to-rural geography. McGee (2008) asserts that with such green spaces, there is the creation of sustainable urban areas, town, and societies.

T- 5 The Urban Centre Zone

The zone comprises of higher density mixed use urban functions. The area or region provide for urban blocks and buildings. Such include retail, offices, apartments and public buildings. In this T5 zone, there are no setbacks and in case they are present, they tend to be minimal. Buildings are proximal to the generally wide sidewalks. Moreover, the zone is characterized by a tight street network that form small blocks. Therefore, the densely populated streets tend to be busy, and they are clearly different from rural regions i.e. T1 or rural area. In other words, there is high connectivity in the urban centre zone. In most cities, the urban centre represents the location of traditional downtowns. The urban centre zone may also have steady street tree planting and parks. Therefore, there is evidence of protected open spaces and public plazas in the region.

T- 6 The Urban Core Zone

The zone represents the region with the highest population density and in some cities may have a high-rise skyline. The zone is characterised by the highest intensity as well as the diversity of uses. In the urban core zone, buildings are designed on notably wide sidewalks. Similarly, there is commendable street connectivity. In most urban regions, the zone has numerous civic buildings that are particularly important to individuals in the region. As Duany, Sorlien, and Wright (2005) notes, the urban core zone may have features such as larger blocks and streets with steady street tree planting and vegetation coverage in the city. Therefore, urban designers also focus on creating green spaces in these regions for purposes of improving the sustainability of urban life. It is pertinent to point out that only large city and towns have such a zone i.e.an Urban Core Zone (Bohl and Plater-Zyberk, 2006).

Other key aspects in the description of the Transect as a design tool are the Civic Zone and Special Districts. According to Duany, Sorlien, and Wright (2005), Civic Zone consists of Civic Buildings and Civic Spaces that serve all the Transect Zones. On the other side, Special Districts are part of areas with blocks and buildings that cannot fit in one or more of the six Transect Zones, as discussed above. The exclusion is due to the specialized function, configuration or disposition of the buildings (Bohl and Plater-Zyberk, 2006). Bohl and Plater-Zyberk (2006) further note that the special purpose districts or areas include airports, institutional campuses and heavy industrial regions.

Significance of the Transect as a Design Tool

When considered in combination with modern-day compensatory overlays, such as telecommunication infrastructure, the internet and emerging mobility networks, the approach of the transect model remains essential for remediating ecological stability and community cohesion in a given geographic context. The rural-urban transect method has undergone significant advances in relation to the conventional ground plan master planning options. In particular, in fragmented ecologies requiring reconstruction the sectional transect method takes into account topographic and environmental variations. This makes it more suitable than the two-dimensional plan option in controlling and guiding urban morphology via different types of environmental and functional correlations, and the embedding of design ideas in section (Behzadfar, Alalhesabi, and Amirhodaei,

2017). For example, Behzadfar, Alalhesabi, and Amirhodaei (2017) explain that the transect method covers and supports types of design approaches and theories that incorporate and take advantage of ecology in the process of developing sustainable urban regions. Previous models for analysis and design of urban areas failed to support the ecological design and the acknowledgement of gradient transitions of rural to urban areas. As presented in the preceding section, the New Urbanists successfully attempt endeavours of differentiating human settlements which is a key addition in this transect approach (Duany, Sorlien, and Wright, 2005; Bohland Plater-Zyberk, 2006).

The transect approach can be explained using two major concepts, namely, ecological and geographical transect. On one side, some studies views transect as a "geographical cross-section of the region" and this portrays the sequence of the various environment types. On the other side, Transect is the method, a design tool, and technique of sampling, analysis and inventory in research on biology, ecology and other related science disciplines (Behzadfar, Alalhesabi, and Amirhodaei, 2017). Besides, the approach aids in accomplishing a detailed view of the environment and also assists in gaining an estimation and evaluation of the ecologies and cultures found in regional scale sections. Urbanists gain a perception of the setting, region or environment by noting the patterns and elements that differentiate one habitat from the other (Duany, Sorlien, and Wright 2005). By applying the transect method, an identification of a spectrum of the sequence found in the environment is also pertinent. Additionally, the detection of broken environmental continuities and fragmented community networks assists designers in making choices for design interventions and the embedding of new planning and remedial ideas for territories.

The transect method can analyse the specifications of natural resources and the pollution found in the environment, for example, minerals, chemical compounds and contamination in soil between the high-risk and the low-risk contaminated areas. In this way, the method assists in checking the extent and impact of the acute risk polluted areas relative to potential land use functions. For example, design interventions may propose to relocate sensitive land use functions, such as schools and hospitals, away from highly polluted soil or water. The complete removal of contaminated soil based on cross-sectional ground studies may be part of a geotechnical design intervention. The design and planning of rural-urban regions can contribute to the reduction of environmental pollution. For instance, advanced transect ecologies, i.e. similar in specifications T4, T5, and T6, provide for green spaces, public parks, and fosters the development of a green amenity network, cycling and walking paths (Bohl and Plater-Zyberk, 2006). In addition, through the promotion of shared transport systems, renewable energy generation, small-scale urban farming and food production at the local level (Young, 2017), environmental issues such as climate change can be addressed when territories are designed using the Transect method. Similarly, food production at the local level could to some extent deal with the issues of food insecurity and poverty in peripheral areas outside the metropolitan agglomerations. According to Deal (2017), different transect modules offer general guidance on types of food production and local food systems that can be sustained in every transect ecology type.

It is pertinent to point out that rural-urban transect can be perceived in terms of the interaction between different elements in dispersed settlement structures. The elements include specifications of the physical environment, forms of land use and the spatial relationships between land, land use types, vegetation, and transportation networks in the region (Behzadfar, Alalhesabi, and Amirhodaei 2017). Understanding such relationships helps compare areas along the cross-section and with other cross-sections or regions (Bohl and Plater-Zyberk, 2006). The rural-urban approach can also analyse the density and abundance of the above elements in line with the cross-section, for example, urban designers and analysts can study the inventory of forest and arable land in a certain rural region or mountain areas. As Bohl and Plater-Zyberk (2006) note, many planners, i.e. urban planners, architects, and transportation planners currently embrace the ideas of the Transect design tool. The planners use the transect method as a formal framework or on an informal basis for purposes of embedding new knowledge in community planning, state-of-the-art technologies, plans to transform settlements and also to organise types of dwellings, farm land, industrial sites, scenic areas for tourism, roads, and other physical environmental aspects (Bohl and Plater-Zyberk,2006).

Conclusion

Ultimately, to overcome a rigid and artificial segregation into different transect zones as suggested by the NUM, the rethinking of the section as a series of interdependent ecology types with fluid boundaries is more adequate. Considering the transitions between geographic units are fluid, the concept of 'evolving ecologies' along a section line instead of 'zones' avoids the making of boundaries. The Transect, which evidently has a long standing history, is particularly applicable in the design and analysis of territories. This discussion considers the Transect as a design tool for scenario constructions. The methodology has a long history in Europe and North American planning discipline as compared to urban planning in China. However, literature detailing the gradient continuum in rural-urban territories is growing steadily, making transect as a design tool receive increased attention and application in design practice. This is, in part, due to the fact that practices in urban planning are operating globally and therefore sharing knowledge.

The transect as a 'smooth' section line, as it is represented in the Geddes Valley section is increasingly difficult to find or to reconstruct in environmentally fragmented territories. For instance, the introduction of transport infrastructural networks, water transfer projects and New Geography networks have created both disconnections and new channels of connectivity at the regional scale.

Hence, the transect as a design tool and geographic unit remains instrumental for reinstating continuities in divided territories. The transect method can be utilized for designing new linkages and overlaps between localized environmental, cultural or economic ecologies along regional scale section lines. By designing-in new linkages and overlaps between ecologies the potential of new synergies in human settlements can continue to evolve.

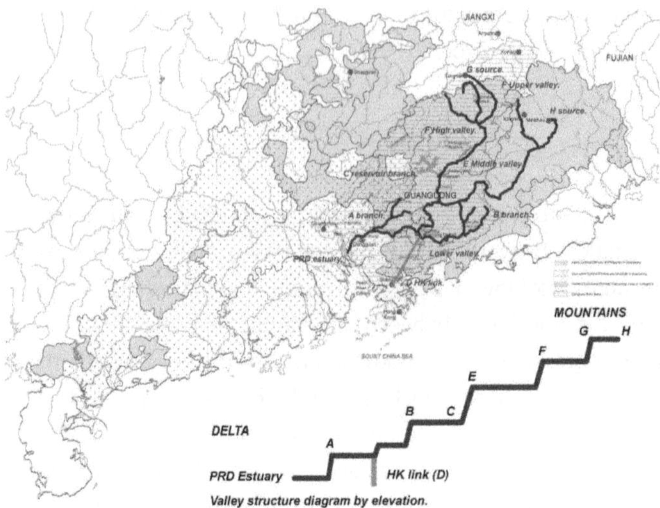

Fig. 4.08: 'A Geddes Valley Section variant' example, elaborated with Prof. David Grahame Shane. This stepping valley diagram explains the structure of the hydrological system of the Donjiang River Basin in its larger regional context. The diagram is a way to organize and analyze research work of the different ascending environments, cultures and ecologies in the river basin. Due to the large scale of the basin territory and its dependent ecologies it is difficult to capture and describe all of the different environments found in this larger scale stepping valley section. In this research project I chose to concentrate on the remote upstream region as it is poorer than the lower zones of basin closer to the mega-city valley. Image generated by the author, 2019.

Recent large-scale urban planning projects in China are adopting regional design concepts, including influences from Geddes, Garden City Movement and Landscape Urbanism and Ecological Urbanism. See project examples from 2014 to 2020 provided in the following section 4.2.1 – 4.2.6. Several of the projects work with concepts acknowledging different types of regional spaces. This part notes the presence of different regions, i.e. rural, corridor, and metropolitan regions. These are similar to the six varying transect 'ecologies' instead of 'zones'. and they can be named natural, rural, rural-urban, peri- urban, urban centre and urban core zone respectively. Civic functions and specialized districts are the other key urban functions that do not fit into any of the six categories.

The transect approach is a significant design tool considering how it has advanced in relation to the ground plan-based master planning option. Precisely, the transect design tool allows to transform and preserve rural-urban morphology through establishing new relationships in sections and also by embedding cutting-edge technologies, and blending them with existing ecologies into design proposals to reactivate distressed territories. The method can also analyse the different specifications of ecological capital and the ongoing environmental degradation. For example, rural-to-urban transect ecologies promote the continuity and extension of ecological corridors, and fosters the development of environmentally friendly mobility. This can increase the resilience and cohesion of different transect ecologies. The transect method is also used as a formal framework to test scenarios for alternative urbanities in remote mountain territories. Last but not least, it can be used as a model to structure large-scale regional research (as examples see Fig. 3.46-3.48; and a Geddes variant in Fig. 4.08). Moreover, the transect method is an effective method to organize designs proposals according to on site eco-system continuities, and also to synthesize types of land use functions, transportation networks, embedding of community spaces, environmental and welfare infrastructure, farmland preservation and food production sites, and the protection of environmental capital for future generations.

4.2. Recent projects influenced by Geddes & the Garden City Movement in China:

Ecology – focused, Urban regeneration and urban-rural planning projects in China.

The following project examples are recent large-scale urban and regional project conceived since the National New Type Urbanization Plan 2014-2020 (NNTUP).

In Summary the objectives of the NNTUP are:

1_Rural to Urban: Transferring population bases from rural environments to cities and towns.
2_Improved city planning: Optimize the layout and form of urban areas.
3_Environmental Sustainability: Enhance sustainable development.
4_Better integrated Society: Properly integrate rural and urban developments to improve living standards outside existing city areas.
5_Reform of Social institutions: improve efficiency of social institutions that will manage urban development. This will include reforming the Hukou system. (Sources: NNTUP, KPMG_2014).

The six contemporary project examples each can be classified into the below project themes:

- Design with Nature: Xingyang University Masterplan, Ecology and Community cohesion. By Sasaki.
- Ecology Restoration in the City: Zhangjiabang urban park: Shanghai, China. By Sasaki.
- Garden city movement: Lingang Ring Masterplan - Shanghai, China. By Sasaki.
- Productive Garden City: Blending Modern Agricultural Ecosystem and Urbanisation. By Sasaki.
- Ecology Restoration in the City and linking together the ecological network: Urban Green Infrastructure reconstruction: NingboEco Corridor. By SWA Group.
- Mountain valley urbanization: Jianjiang Valley Guide Plan, China. By Agence Ter.

4.2.1. Design with Nature: Xingyang University Masterplan, Ecology and Community cohesion.

As designed and elaborated by Sasaki, the master planners, ecological design and designing with natural landscape features drive the master planning project for the new Xinyang University campus. The blending of learning and academic communities with an environment wealthy in ecological resources is one of the main aspirations for growing a sustainability conscious and socially aware university community. The very unique natural landscapes and outdoor spaces have been considered for the development of the project.

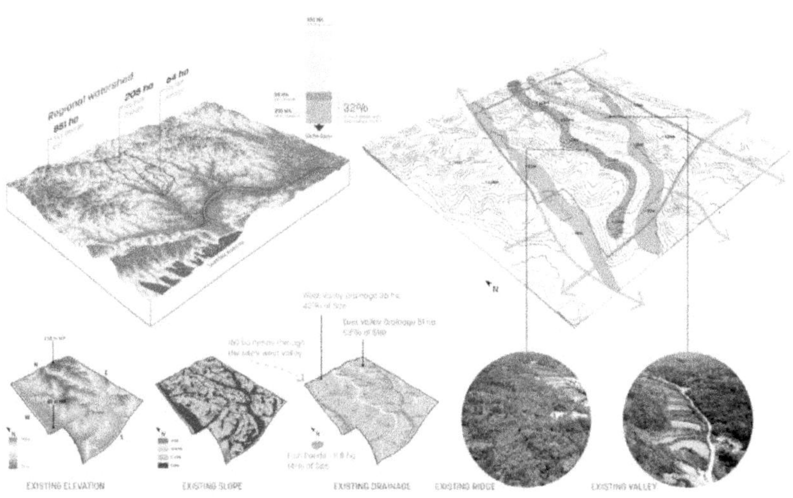

Fig. 4.08a: The 86.7 hectares master plan for in Xinyang City, Henan Province, China was conceived in 2020 for the Xinyang University. As a starting point for the masterplan design the topograpgy and hydrological network is the basis for this scheme. Source: Sasaki, 2020.

Design with Nature: Xingyang University Masterplan, China.

The site topography with valleys and mountain ridges are an important spatial capital of the master plan. The natural spaces form recreational spaces and outdoor student amenity areas for education. The valleys also carry eco-system services including storm water management and habitats for bio-diversity.

Fig. 4.08b: The hilly terrain, valley spaces, mixed forests and post-agricultural landscapes have been incorporated into the design. Source: Sasaki, 2020.

Design with Nature: Xing yang University Masterplan, China.

The masterplan has been designed to support a pedestrian-oriented environment. All the University Faculty buildings and key amenities can be reached within five-minutes walking distance relative to each facility. A well linked group of community, sports and recreational buildings constitute the infrastructure of the campus buildings. The Student dormitories are nestled in clusters into the landscape terrain.

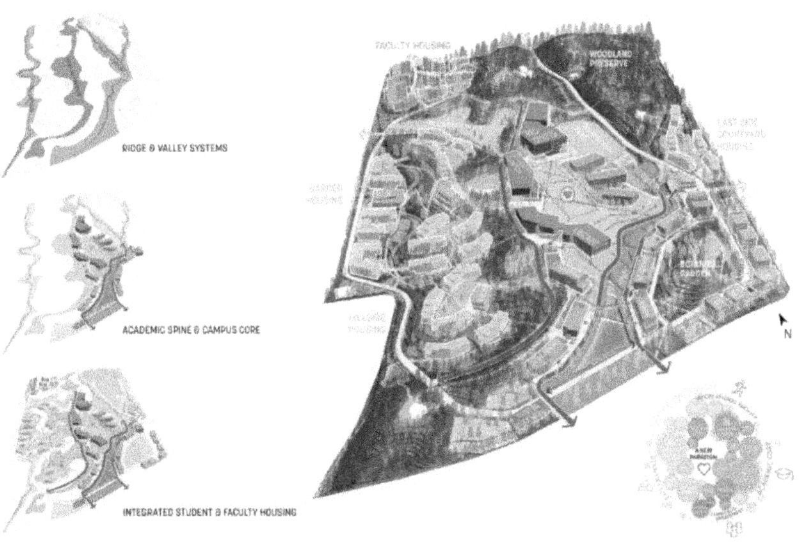

Fig. 4.08c: The valleys of the landscape are the spatial and organizing feature accommodating the social amenities of the campus master plan. Source: Sasaki, 2020.

Design with Nature: Xing yang University Masterplan, China.

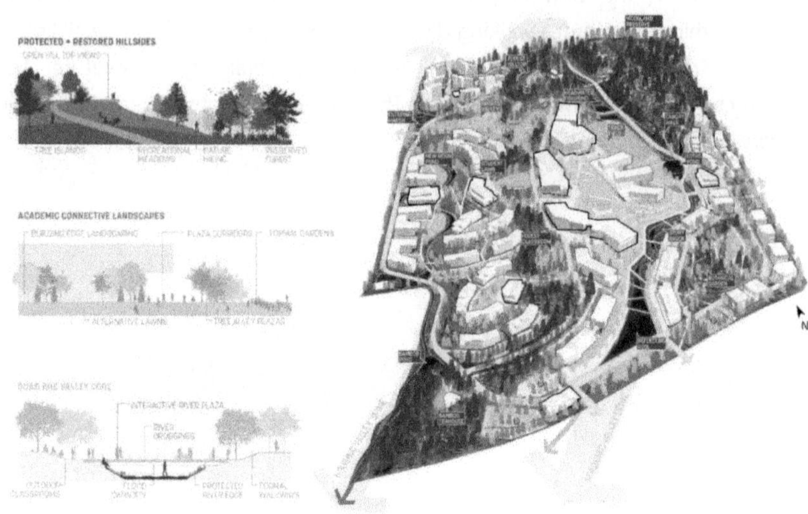

Fig. 4.08d: Preserving the eco-system services and designing with nature has been part of the planning process. Source: Sasaki, 2020.

Design with Nature: Xingyang University Masterplan, China.

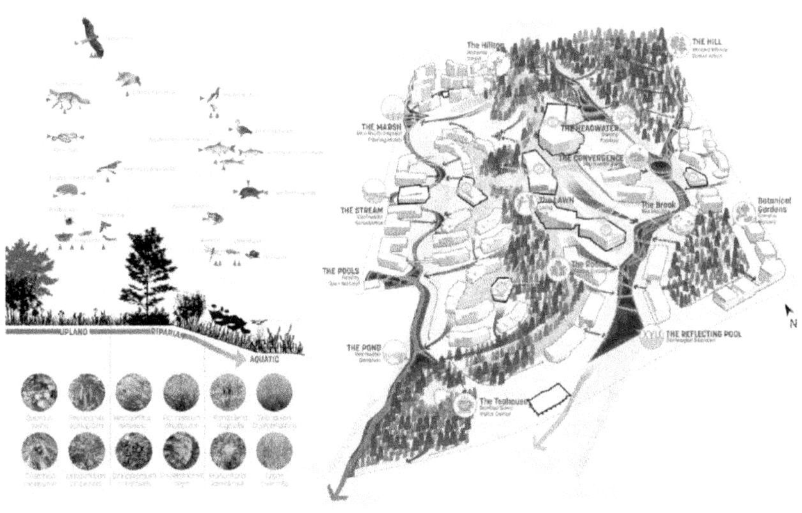

Fig. 4.08e: The natural site drainage has been considered to avoid disrupting the existing hydrological system. Source: Sasaki, 2020.

Design with Nature: Xingyang University Masterplan, China.

State-of-the-art sustainable building design and planning strategies have been deployed for this project. The holistic approach includes strategies for energy efficiency, waste recycling, preservation of topsoil and local climate considerations as part of design strategies. Maximizing opportunities for sustainable design measures such as natural ventilation, capturing winter solar gain, the installation of photovoltaic panels and solar water heaters on the roofs included as part of the design interventions. Careful management of waste and composting of bio-degradable waste from kitchens is re-used as organic matter to provide nutrients to the soil on site. To reduce the impact of the construction activities and damage to the landscape, the top-soil will be stored and then re-instated on site after the construction works.

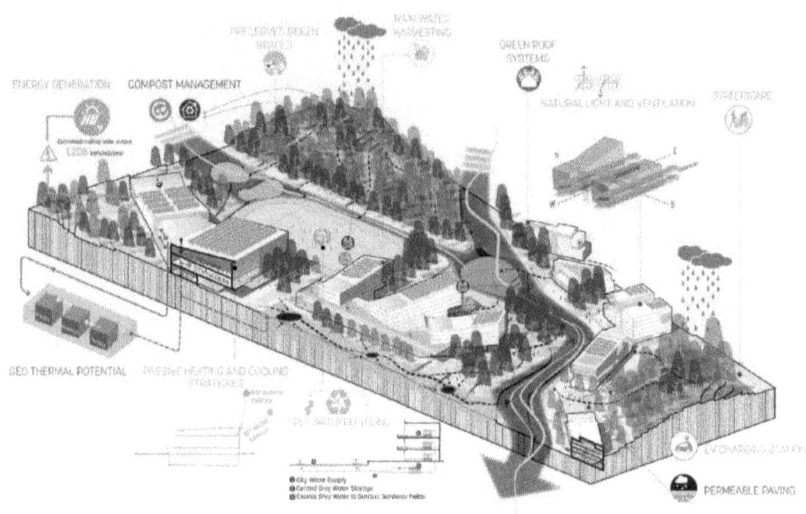

Fig. 4.08f: Several ecological design strategies have been deployed including clean energy generation, geothermal, rain water harvesting, grey water recycling, compost management, natural light and ventilation, passive heating and cooling strategies, EV charging stations, green roofs and permeable paving. Source: Sasaki, 2020.

Design with Nature: Xing yang University Masterplan, China.

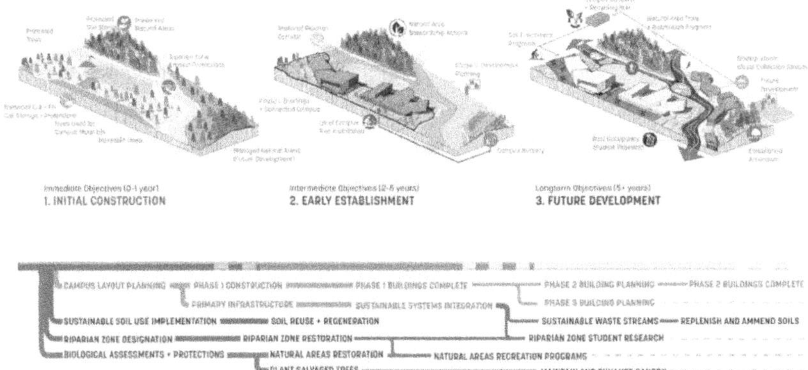

Fig. 4.08g: To manage the environmental impact of the development and construction activity, the masterplan development has been phased into three stages to preserve the natural areas, trees, and riparian corridor erosion protection has been planned as part of the phasing considerations. Source: Sasaki, 2020.

4.2.2. Ecology Restoration in the City: Zhangjiabang urban park: Shanghai, China.

As part of a Green Ecological Network, Shanghai has eight 'green-wedges' planned. As explained by Sasaki, Zhangjiabang park is one of the green infrastructure provisions for the city to counteract the loss of green open spaces caused by urbanization. On an 843 hectares site, Zhangjiabang is the largest park for the city and intends to restore a habitat with wetland and woodland zones to improve the environment and enable eco-system services to be revitalized.

Fig. 4.08h: The aim of this 843 hectares urban green infrastructure project in Shanghai is to restore the ecology in the urban area, improve its micro-climate and to re-connect green amenities with the urban fabric. Source: Sasaki, 2015.

Ecology Restoration in the City: Zhangjiabang urban park: Shanghai, China.

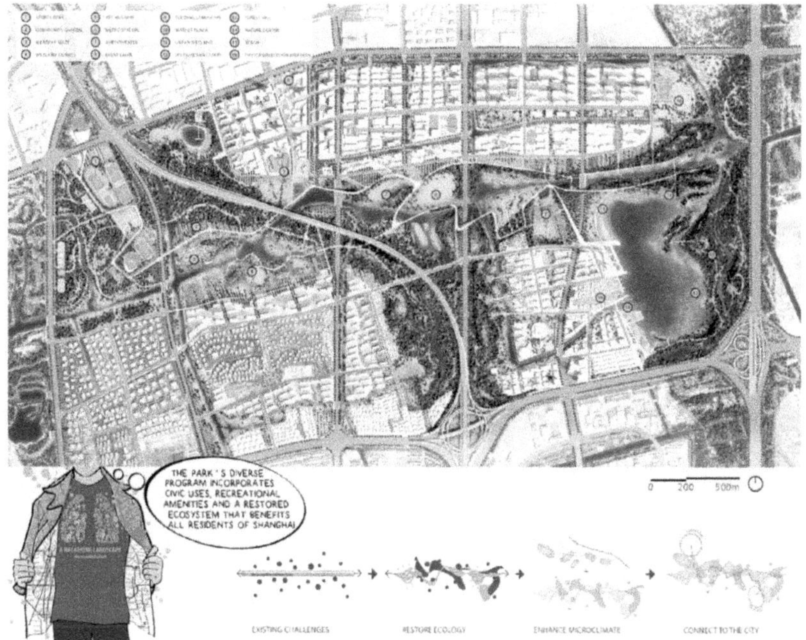

Fig. 4.08i: The three main ideas guiding the development of the urban park are the restoration of ecology, enhancement of the microclimate and reconnecting the part to the urban fabric. Source: Sasaki, 2015.

Ecology Restoration in the City: Zhangjiabang urban park: Shanghai, China.

The landscape master plan focuses on the current canal and its tributaries, improving the water quality and restoring the biodiversity of the aquatic ecosystems in the urban area. In addition, air quality and thermal comfort are improved by careful spatial arrangements using landforms, planting techniques, water bodies, and prevailing winds to create microclimates that contrast with the stifling urban heat island of Shanghai. Also integrated is adjacent development which reconnects the urban fabric on either side of the park. New transport routes connecting Zhangjiabang to Shanghai's inner urban areas provide urban entry points to the site, which create an opportunity for mixed-use destinations due by their easy access to the park.

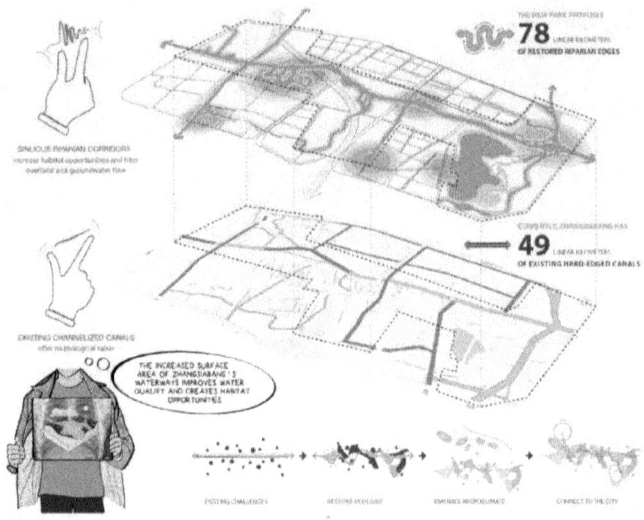

Fig. 4.08j: Existing hard edge canals with no biological value are restored into riparian ecology corridors. The remediated riparian corridors form habitats for bio-diversity, and purify surface and groundwater on site. Source: Sasaki, 2015.

Ecology Restoration in the City: Zhangjiabang urban park: Shanghai, China.

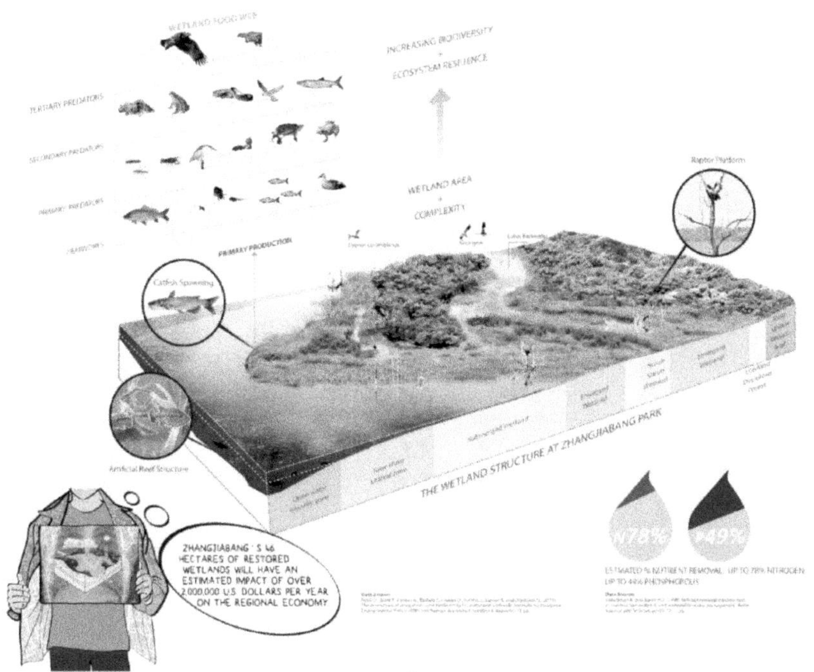

Fig. 4.08k: The following Wetland zones have been considered for restructuring the riparian edges: Evergreen broad leaf zone, Emergent wetlands, Shrub wetland, Emergent - Submerged wetlands zone, Near-shore littoral zone, open water limnetic zone. The complexity and diversification of the wetland zone contributes to eco-system resilience. The eco-system resilience also supports biodiversity, which then cultivates a diversified food web for a self-supporting, self-sufficient ecology. Source: Sasaki, 2015.

Ecology Restoration in the City: Zhangjiabang urban park: Shanghai, China.

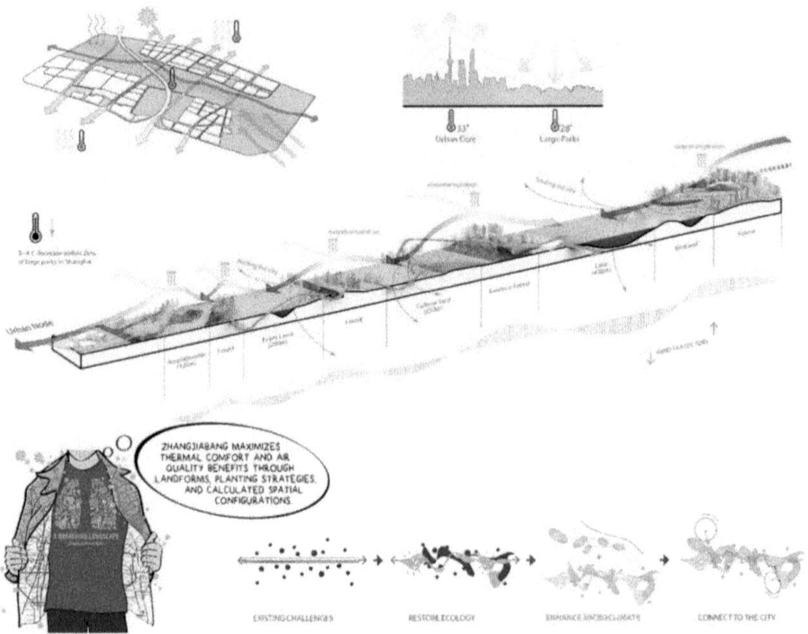

Fig. 4.08l: The new park is also conceived to regulate the urban micro-climate. This is an adaptation strategy for future global warming and the mitigation of the local urban heat island effect. It is expected that a temperature decrease of 3-4 degree Celsius within 2km of large parks can be achieved. Source: Sasaki, 2015.

Ecology Restoration in the City: Zhangjiabang urban park: Shanghai, China.

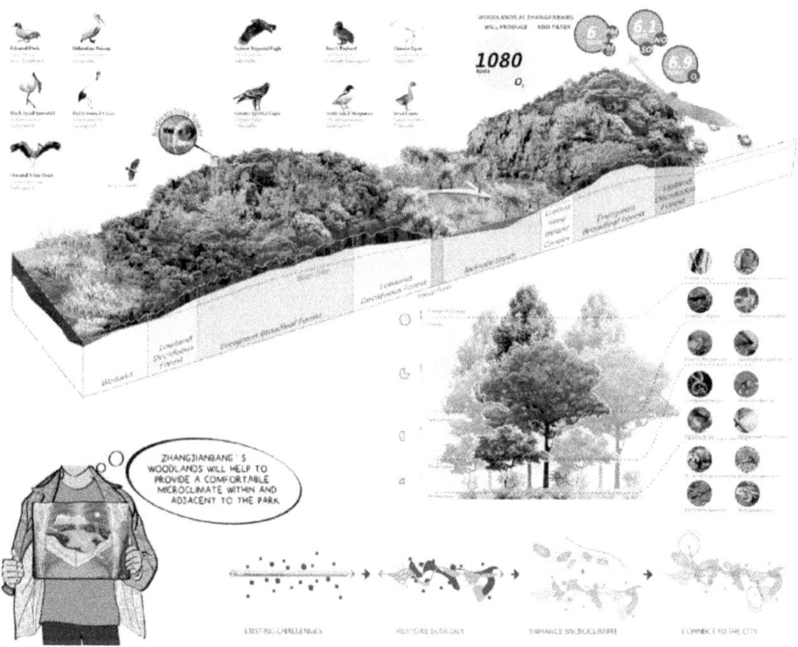

Fig. 4.08m: Air purification and pollutant absorption is one of the eco-system services the park will provide to the urban area and communities. The parks woodlands have the capacity to release 1080 tons of Oxygen. The park will absorb 6 tons of PM2.5/PM10, 6.1 tons of Sulphur Monoxide (SO), 6.1 tons of Nitric Oxide (NO), 6.9 tons of Trioxygen (Ozone). Source: Sasaki, 2015.

4.2.3. Garden city movement: Lingang Ring Masterplan - Shanghai, China.

This urban regeneration project by Sasaki is a central part of the local park spaces program and an essential contribution to the larger Dishui Lake ecosystem. The park is closely connected to the surrounding communities and open spaces, providing local leisure amenities and wildlife habitat areas, linking people's lives, jobs and space for recreation to the regional ecological environment. While considering the open space network improvements already made in this urban area, several habitats are planned for the new green infrastructure, including the planting of trees, meadows, lawns, creation of farming areas, ponds, and public plazas. The remainder of the regeneration site, which has not yet been remedied, has a high degree of salt in the soil. As the construction of the adjacent land will take a long time to accomplish, the aim is to remediate the soil through sustainable agricultural and pastoral practice over time. The idea is to gradually eliminate the saline content, raise the nutrient level, and incrementally plant and mix the plant species with tree planted areas. Large areas of wetland outside the park site will be integrated into the desalination cycle, which will continue to act as a barrier to prevent seawater from flooding the water-to-land transition zones and helps to attenuate large storms and waves. Although Lingang will take several years to develop, the park will evolve its shape, transforming in phases over time while the adjacent developments progress. The park is expected to contribute more to ecological restoration and remediation during the early stages of growth. In contrast, the park landscape should become more mature and offer a very diverse, urban green destination and recreational amenities in Lingang, when most developments are completed

Fig. 4.08n: Lingang's 546-hectare Green Ring Park - Lingang New City Masterplan. Source: Sasaki, 2015.

Garden city movement: Lingang Ring Masterplan - Shanghai, China.

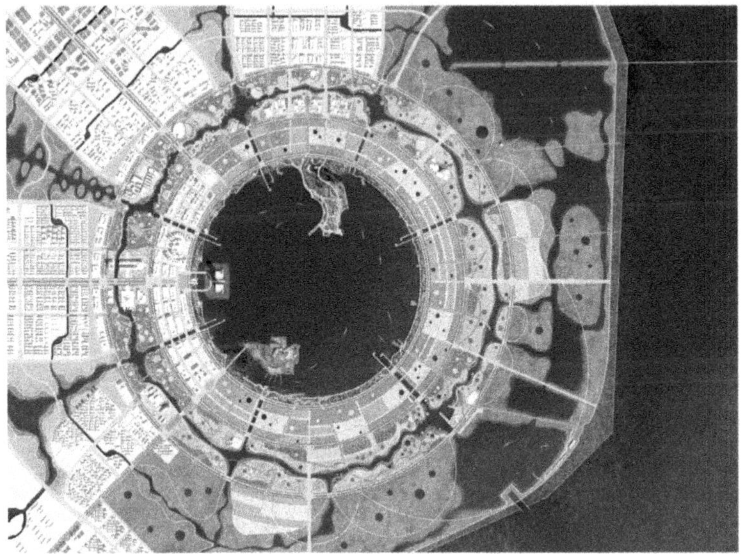

Fig. 4.08o: The radial organization of the masterplan synthesizes urban development, agricultural land use, green infrastructure, tourism and recreational destinations. The development is intended for 800,000 future inhabitants. Source: Sasaki, 2015.

4.2.4. Productive Garden City: Blending Modern Agricultural Ecosystem and Urbanization.

Garden City movement and Productivity.

South Taihu Lake Master plan. As Sasaki, the master planner, explains this project, food security is increasingly being undermined due to the expansion of urban land in China. China lost to urbanization over 123,000 square kilometers of agricultural land from 1997 to 2008. In order to challenge the idea that the consequence of economic development is that agricultural land becomes reduced for urban development, the master plan for South Taihu Lake exemplifies how agricultural land can adapt and contribute value to the local communities. Therefore, the project includes a variety of elements comprising a modern agricultural ecosystem, including food processing, as well as research and development for agricultural industries. In addition, the project addresses the sustainable re-use of a historic village featuring local farms for table dining and the ecological regeneration of a once highly deteriorated site. The design focuses on a productive landscape that gives value through the creation of a novel business model to attract regional investment. For agriculture and open space protection, the scheme retains over 75 per cent of the site, preserving it for food processing, agricultural research and development and urban green infrastructure. Agricultural business ecologies concentrate on various sectors that are either growing or are already booming in the area. Orchards and other choose-your-own seasonal crops from neighbouring metropolitan areas such as Shanghai and Hangzhou attract an emerging agri-tourism industry. Organic, locally grown produce farms are used to supply high-quality products to local hotels and restaurants, including those in the restored historic village on the farms and textile industries. As part of the proposed master plan, the pilot projects provide new innovative crops for research and development from the existing biotech industries in the region.

Fig. 4.08p: South Taihu Lake master plan is a 1,312 hectares planning and urban design project in Huzhou, China. Source: Sasaki, 2014.

Productive Garden City:

Blending Modern Agricultural Ecosystem and Urbanization.

Cultivated farm land on site is made accessible to introduce tourists to the local food production, the agricultural supply chain, and the multitude of business ecologies that depend on raw materials for crops. New development parcels are arranged around established villages, incorporating traditional village fabrics and creating a special mix of contemporary architecture within retained historic building stock.

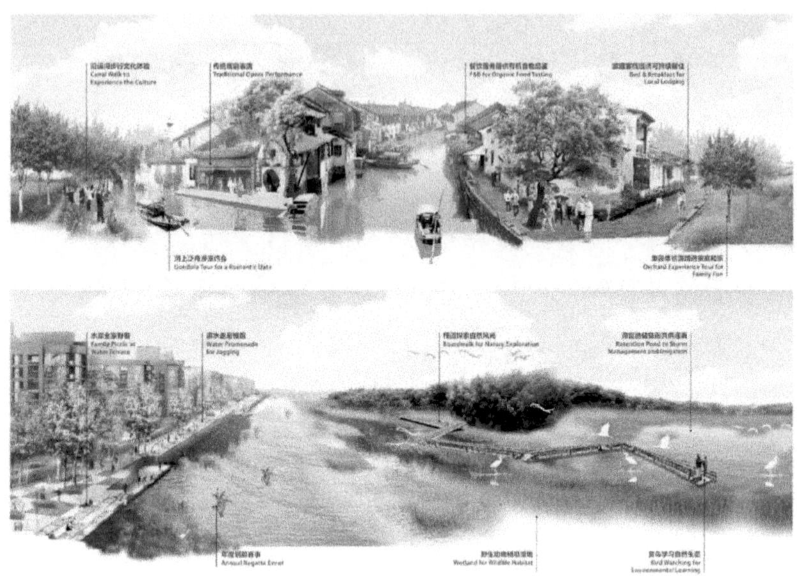

Fig. 4.08q: The masterplan integrates traditional villages houses and water sports amenities as part of the productive infrastructure to attract tourism to the local economic activities. Source: Sasaki, 2014.

Productive Garden City: Blending Modern Agricultural Ecosystem and Urbanization.

A very thorough on-site environmental study showed that approximately sixty-five per cent of the one-thousand-three-hundred and twelve hectare project site is below Taihu Lake's high water level. Despite the existing levees protecting the site, seasonal floods are still an environmental risk. The master planning proposal plans for a water-receiving landscape – including restored wetlands – that can help to purify the water and reduce the risk of flooding. Into the existing pattern of cultivated land, a system of interconnected patches of wetland is integrated.

The wetland patches are arranged to correspond to the natural water flows on the site that drain to Taihu Lake. This interwoven network of wetlands provides a buffer of the nutrients from agricultural activity, enhances the water quality, and supports biodiversity for a variety of migratory birds and local species.

Fig. 4.08r: Agri-cultural tourism parks and food security for the local population are combined with green infrastructure to maintain the local eco-system stability. Source: Sasaki, 2014.

4.2.5. Ecology Restoration in the City and linking together the ecological network.

Urban Green Infrastructure reconstruction: Ningbo Eco Corridor.

The masterplan developed and, as explained by SWA, the metropolitan Ningbo Eco-Corridor is 3.3 km long and has an area of 250 acres. The project regenerates a former agricultural plain that had been taken over by industrial use into a new urban green infrastructure. Situated within the Yangtze River Delta on the Chinese coast, Ningbo is one of China's largest towns, with 3,616 square miles and a population of 5,43 million inhabitants. Ningbo, which in Chinese means "tranquil waters," is an important port for international trade. The growth of this region prompted a plan to create Ningbo as a larger metropolitan area of economic and environmental significance and prepare future development for an ecological approach to the regeneration of the city. To provide Eco-system functions, the scheme conceived as a 'living filter' uses the environmental and landscape design techniques of hydrology, vegetation and topography to provide recreation, education and cultural facilities for Ningbo East New City. Every planning area planned proposes a sustainable ecological approach by treating and recycling run-off surface water, capturing solar energy, and managing the local urban heat effect. In addition, the scheme incorporates a sustainable hydrological system that allows water to be efficiently recycled, stored and redistributed for resourceful consumption. The project's goal is to cultivate awareness for local inhabitants and their climate, promote a better relationship between people and their water resources, and opportunities for sustainability and ecology-conscious education. The Ningbo Eco-Corridor improves public health, contributes to the quality of life, and raises property values of local and neighbouring communities by connecting the ecological network in this area. Last but not least, the project encourages other cities in the area to support ecological design strategies and master planning ideas for their future urban transformations.

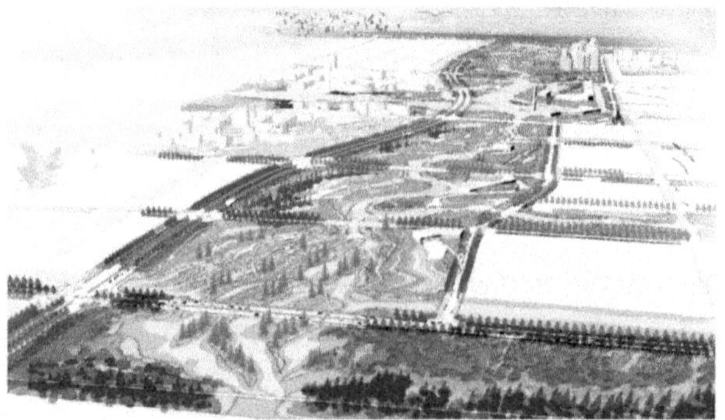

Fig. 4.08s: The eco-corridor project was initiated for the Ningbo Planning Bureau – East New Town Development Committee in 2014. Source: SWA Group, 2014.

Ecology Restoration in the City and linking together the ecological network.

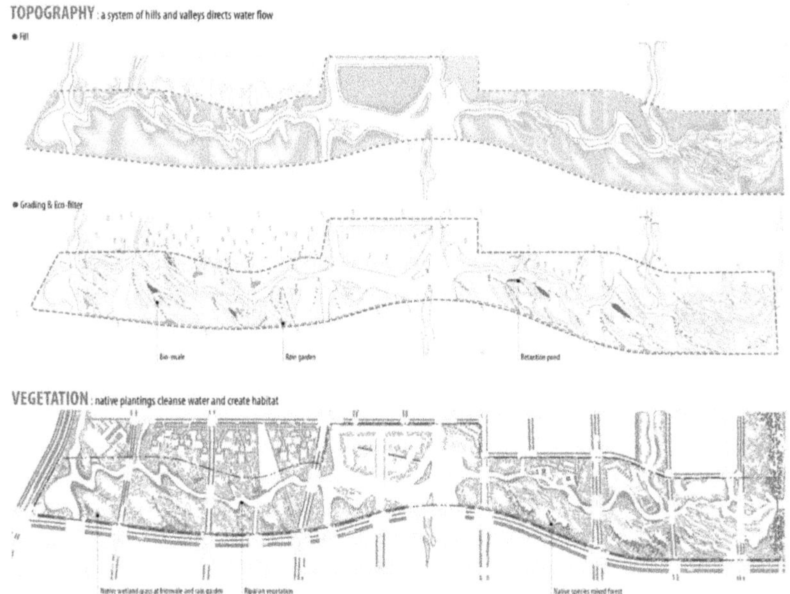

Fig. 4.08t: The 'living filter' – eco-corridor project utilizes the topography and ridge lines of the site to purify, direct water flow and collect rain water. The landscape grading strategies and vegetation planting are used for water cleaning and habitat creation. Source: SWA Group, 2014.

Ecology Restoration in the City and linking together the ecological network.

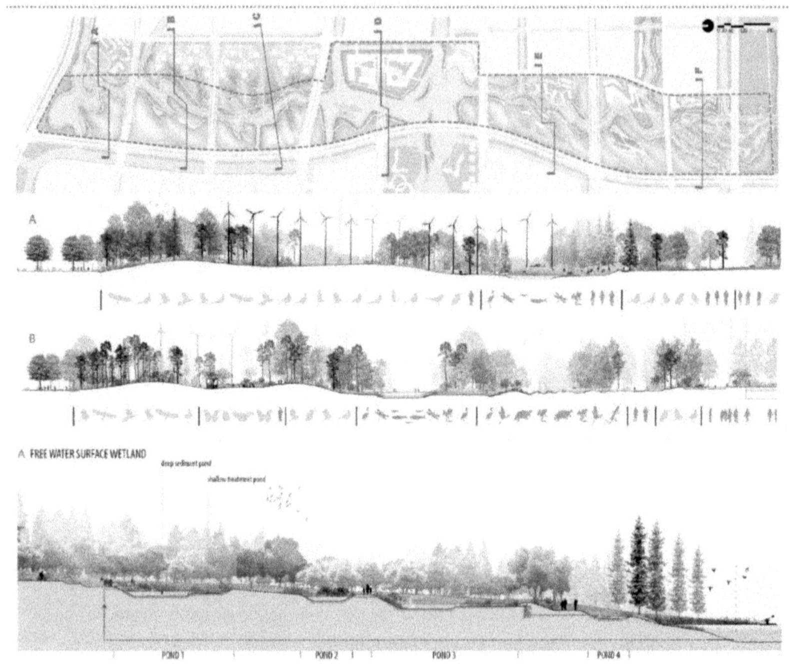

Fig. 4.08u: The riparian wetland zones are structured into cascading ponds collecting and filtering surface and rain water. Source: SWA Group, 2014.

Ecology Restoration in the City and linking together the ecological network.

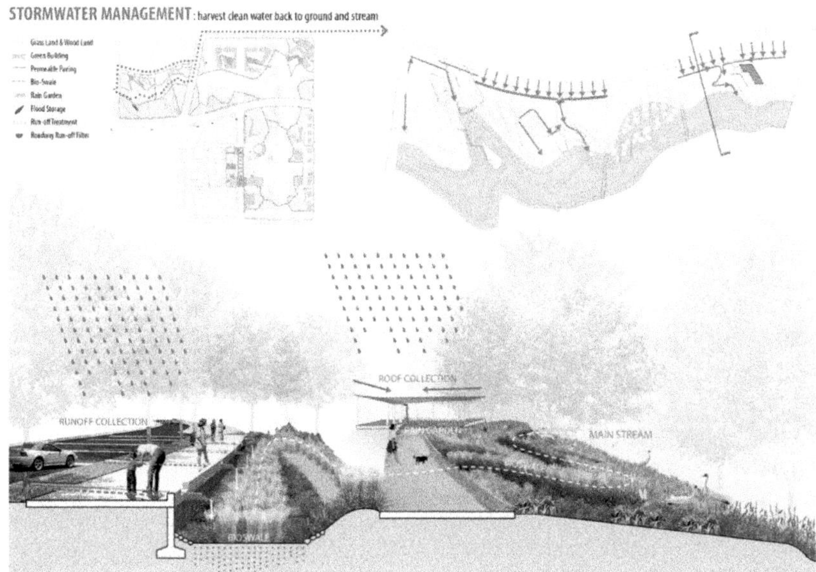

Fig. 4.08v: The following design interventions have been incorporated into the design for the purpose of storm water management: Grassland and woodland areas, bio swale zones, permeable paving areas, rain garden water collection, and surface water collection strategies, water run-off filters and treatment details. The landscaping design includes flood storage capacities and the channeling of clean water back to the ground and stream. Source: SWA Group, 2014.

Ecology Restoration in the City and linking together the ecological network.

Fig. 4.08w: The completed eco-corridor project not only serves the local communities in the city but also raises awareness of the environmental green infrastructure needed and is an example to follow for future urban regeneration projects in China. Source: SWA Group, 2014.

4.2.6. Mountain valley urbanization: Jianjiang Valley Guide Plan, China.

The Jianjiang Valley Strategic Master Plan by Agence Ter consists of three main strategies. The designers outline the following design approach:

- Reorganization of access and mobility in the valley between settlements
- Community planning as well as protection of natural habitats and resources.
- Construction of the seven new destinations in the valley to support sustainable development.

The valley can be understood through its two sloped topography and the following three distinct topographical level. The river bed is the first topographic element; the intermediate slops are the second, and the crest, a unique characteristic of this valley, is the third element. Each topographical level is interrelated with newly embedded architectural design projects into this territory. Six main villages and three smaller villages make up the valley network. For the masterplanners, one of the challenges was how to enable settlement densification, the integration of infrastructure and development while avoiding degradation of the ecological environment in this landscape. In addition, the project proposes interconnecting the villages to reinforce their potential for new synergies and sharing of resources.

- The reinforcing the resilience of each village needs the following:
- Densification of the current settlements by recycling and filling in vacant lots between existing building.
- Enhancing urban-rural transition zones, strengthening of ecology zones, and the promoting of clean sources of energy. To rejuvenate the agricultural livelihood ecologies and farming activities in the valley, the protection of arable land and fertile soil is vital for its inhabitant and the maintaining of biodiversity.

Fig. 4.08x: Jianjiang Valley, Pengzhou. Project area 150 km^2 Source: Agence Ter, 2018.

Mountain valley urbanization: Jianjiang Valley Guide Plan, China.

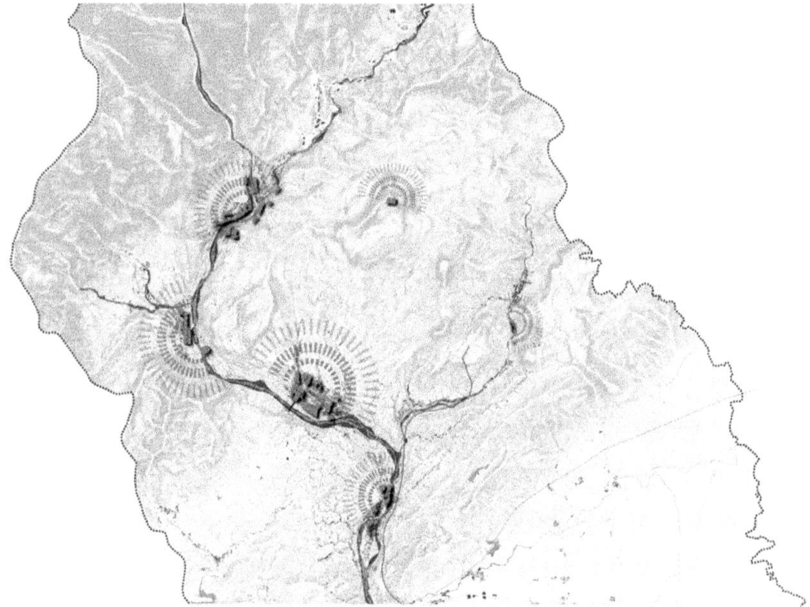

Fig. 4.08y: The Jianjiang Valley masterplan is a study of creating synergies between 6 main villages and 3 smaller settlements. One of the objectives was to densify the network of villages, while preserving a balance of urbanization and maintaining rural areas and its ecology. Source: Agence Ter, 2018.

Mountain valley urbanization: Jianjiang Valley Guide Plan, China.

Fig. 4.08z: The urbanization plan in the mountain valley considers the protection of fertile soil for agricultural activity, and carefully manages the transitions between natural ecology to urban areas, while maximizing opportunities for renewable energy generation. Source: Agence Ter, 2018.

Part Five

Conclusion and future research questions.

5.0. CONCLUSION:

In conclusion, the testing of the transect as a method for regional design problems in Mei County, China in this book has triggered the re-thinking of the transect and valley section for large-scale planning projects. This summary provides an appraisal of the transect as applied in the study sites of remote villages presented in Part Three. Moreover, based on the key lessons learnt from the design studies conducted, this conclusion proposes the revisions required to the transect planning method in the age of an increasing informatization of hybrid rural-urban areas and platform-based 'New Geographies'. ITC supported, post – fordist transactional and social media networks have altered the relationships of people in their physical habitat and linkages with geographically distant regions. Crucial to the revisiting of the transect planning method is to utilize this tool to reconnect divided territories and fractured natural and social ecologies. For a different way of thinking about sustainability-driven opportunities in transect design studies, it is pertinent to abandon binary distinctions between 'natural' and 'anthropogenic' ecologies. The contemporary regional design calls into question the artificial separation of regions into 'city' and 'countryside'. As elaborated in Part Four, the rethinking of the transect as a series of interdependent ecology types with fluid boundaries along a section line instead of 'zones', is a non-binary alternative in line with the notion of environments with no definitive dividing lines. Furthermore, a novel transect approach which considers the unequal relationships between 'center' and 'periphery' of territories, seeks to rebalance the access to information, ecological, social and economic resources, in favor of marginal communities.

Due to the global-scale phenomenon of capitalist urbanization, the long-term problem of ameliorating poverty in mountain regions and the difficulty to develop independent economic ecologies in remote territories through regional planning has become an ever more difficult task for regional planners. On the one hand, regional design for peripheral territories became more challenging as a result of short-term economic cycles. One the other hand, new opportunities for overcoming access to information, education and opportunities for reducing transportation time for people and goods continue to be improved.

In our contemporary world, the reasons which make balanced rural – to –urban development problematic to maintain, are the global competition for resources and increasingly fragmented natural eco-systems. The reassessment of design problems in metropolitan cities and the methods for resolving urban design projects adopt innovative concepts and methodologies more rapidly compared to innovations for regional planning projects. Currently, three examples for pervasive preoccupations in city design would be the implementation of smart city concepts and the use of cloud computing to predict energy use of households, consumer behavior, or crowd movement of people in cities. Perhaps, due to the comparatively lower investment in regional research by governments and private corporations, which are focused on global cities as economic engines, the innovation in regional design methods receives limited attention.

The Geddes valley approach reconnects human livelihood activities in their ecological context and correlates resource extractions in territories with people's occupations (Geddes, 1915). The underlying assumption of the mountains-to-river delta section as a geographic unit, emphasizes a smooth continuity for a regional scale section. By emphasizing continuity, the Geddes valley section consequently prescribes the search for a holistic and seamless approach to regional design. As a design paradigm, it establishes an expectation for designers to design-in the idea of continuity and connectivity into a territorial project. Nevertheless, when studying mountain-to-delta regional sections for example, transects do not necessarily manifest continuity. Specifically, in regions with new transport infrastructural overlays, high-speed rail networks, water-diversion aqueducts and newly developed settlements, divisions, boundaries and discontinuities of ecological systems are revealed. Additionally, the internet was enabled the pursuance of livelihood activities remotely. Hence, this is disconnecting people from the environment they inhabit and places of resource extractions. The contemporary relevance of the Geddes transect paradigm, lies in the reconciliation of humans, their activities and places they live in. It is precisely, this symbiotic ecology which requires reinstatement through regional design which is missing in the currently, prevailing capitalist urbanization phase. Furthermore, Geddes' interest in local cultures which evolve in a geographic context, remains significant especially in an age of globalization and internationalization of culture. As shown in the Mei County study sites of this research, the preservation of local Hakka culture, for example, is paramount for the development of ethnic tourism and the maintaining of a unique cultural identity.

In comparison to the mountains-to-river delta sectional approach by Geddes, the New Urbanist Movement (NUM) concentrates on an urban-to-rural transect approach. Predominantly, the New Urbanist approach to transect planning is a two dimensional plan-based planning system, paying limited attention to topographic features and elevational change of study areas. The NUM transect approach emphasizes an idealized, gradient transition from urban to rural areas. The urban-rural gradient in NUM projects as 'aesthetic representations' are hardly found in North American or European regional planning projects, and much less in Chinese urban-to-rural transects. Due to the representational aspect of the urban-rural gradient, NUM projects become a 'beautification' of suburban development projects and the making of sceneries of urbanity, instead of creating civic urban spaces and genuine urbanization. Nevertheless, as a transect method the NUM and Smartcode by Duany, Sorlien, and Wright (2005) can offer a structured design approach, which as a reference for planning professionals and designers can assist for the rationalization of complex regional contexts. Although, the urban-rural gradient can result in formal pattern making, the aspiration of designing seamless urban-rural transitions is relevant in fragmented territories and ecologies. As presented in the design studies in Part 3.6, a better linking of open green spaces, farmland and parks with settlement areas addresses the need to diminish segregations between built-up settlement areas, and cultivated land and recreational green amenity areas. The design studies for Chiguang town, Longbei town, Changsha town in Fig. 3.97a, Fig. 3.97d, and Fig. 3.97g explore the 'softening' of artificial boundaries by increasing the vegetative coverage on building roofs, walls, the creation of parks on underutilized plots of land between buildings and the planting of trees along roads and the perimeters of plots.

The transect planning method by Geddes and New Urbanism Movement both engage with the material physical assets of a region such as uncultivated land, agricultural plots, forests, hydrological system, buildings, developable land and inhabitants. For both regional development methods the livability of the environment and inhabitability of settlements is a vital consideration for planning projects. In contrast, the regional development approach pursued by Alabama and its Rural Taboo program, targets the development of an on-line retailing network. The development of better road networks and the setting up of goods distribution shop fronts in settlement are the necessary physical hardware. The forming of new linkages with distant cities and locations where products are purchased from or dispatched creates new transactions that can contribute to additional revenue for impoverished mountain villages. Some villages may generate a fortune from a 'Chinese-Rural Dot.com bubble'.

Additionally, a network of Taboo shops and the provision of e-commerce training could take care of urgent issues requiring attention in marginal communities. Such urgent issues include the provision of social services, community centers, health care, provision of education, elderly care, affordable housing, sanitation, environmental remediation, reduction of pollution, investment into green infrastructure and renewable energy technologies. Ultimately, the creation of welfare infrastructure in settlements could create settlements where the well-being of people matters.

In summary, the design studies for Chilung town, Longley town, and Changsha town highlight the need to include the provision of social and environmental design agendas in regional design projects. Both the Geddes valley transect method and the NUM urban-rural gradient concepts are still relevant for rural development in China. The Geddes sectional tool regains significance due to the need to reconcile human livelihood activities in their geographic context. Equally, important is the NUM design system for its attempt to gradually blend urban and rural spaces while considering different human settlement types.

Combined with the embedding of new technologies and ITC platform-based networks, both transect methods remain instrumental tools for addressing the transformation of territories. The hybrid design methods explored in this research underscore the necessity to consider planning problems comprehensively, paying attention to the transactional exchanges of remote territories with Mega-city clusters, while constructing scenarios which promote an independent development of resilient settlements in mountain areas. In particular, the Pearl River Delta (PRD), undergoing large-scale transformation as part of the Greater Bay Strategic Plan, focuses on developing new economic centers within the Mega-city cluster. As the Greater Bay Plan emphasizes the development of cities with specializations within the PRD, a mega-city scale transect approach which captures the River Delta and the Mountain regions in the hinterland could raise more awareness of peripheral territories along a transect. The use of the valley transect method could therefore lead to a more balanced regional development.

Last but not least, the below questions and interests could evolve into future research:

1. A follow-up research into how the ecology, communities and settlements studied in Mei County have been positively or negatively affected by the expansion of e-commerce networks into marginal communities.

2. From the perspective of the spatial design disciplines, how can the architectural and urban design quality in small towns and villages be improved? Can better design standards for community and public buildings be implemented despite the limited resources in remote communities?

3. Can the top-down synergies of Chinese corporations with the community administrations deliver more effective and sustainable development in neglected territories, when compared to sustainable development efforts in other large and populous countries? What are the implications for future urbanization policies in China and other large countries?

In late 2014, when the Alibaba on-line retailing giant experienced a slowdown of growth in Chinese cities, it gradually began to expand its rural customer base. Within a time-span of five years, the rural expansion campaign has attracted international media and research attention. While is it too early to assess the long-terms effects, this nation-wide approach to rural development, despite the current economic downturn in China, will certainly manifest transformations in the way rural communities live and evolve their towns. Hopefully, the quality of the environment, decent housing, health and well-being of people will be priorities driving the next phases of urbanization in China.

* SAR, HK. October, 2021.*

List of Figures:

clusters along five main urbanization corridors or in other words 5+9+6 spatial structure. These are the coastal cities corridor axis, Yangtze River urbanization corridor, Eurasian land bridge urbanization corridor, Harbin-Beijing-Guangzhou urbanization corridor, and Baotou-Kunming urbanization corridor. As such they form a macro-pattern of urbanization for Greater China. Credits: Chuanglin Fang, 2015.

Fig. 1.08: Garbage covered in green protective nets shown as traditional Landscape Paintings reminds the viewer of the nostalgia about the countryside and the issue of environmental degradation in rural areas in China. Credits: Artwork by Yao Lu.

Fig. 1.09: The Geddes Valley section: 'The Notation of Life superimposed on the Valley Region', from Victor Branford and Patrick Geddes, The Coming Polity, Published by Williams & Norgate, London,1917, pp.296.

Fig. 1.10: The image by a Chinese agricultural management company or so called 'dragonhead company' is an idealized illustration of a consolidated agricultural unit. The operational unit consists of 1000mu collective farms (1mu=666.7m2). The aim of agricultural consolidation by dragon head companies is the efficient cultivation of fields and breeding of live stock through the implementation of supporting technologies, integration of land resources, unified management, and promotion of agricultural technology innovation. In this way, the unit plants oat and tea crops and feed 1000 sheep (including 200 basic ewes, 800 lambs and fattening sheep) on more than 1,000 mu of land. This operating unit can make the farmer feed 1000 sheep, plant 150,000 kg of oat or 250,000 kg of potatoes to create a direct economic benefit of 350,000 yuan/pa. Source: San Zhu Liang Group.

Fig. 1.11: Distortions in the Urbanization System: Depending on which sources are looked at the floating population is estimated between 120 to 274 million people in 2015. Due to the informal and temporary settlement of migrant workers an exact figure is difficult to derive. By 2020, the floating population may become up to 291 million people. Various Sources: Paulson Institute, SCMP, China Daily, China Economic Watch / Urbanization and Economic Growth China. Nicholas Borst (PIIE), Sept. 2012.

Fig. 1.12: The phenomenon of a middleclass in China is a relatively new concept. It is a society in the making, defining how to find livelihood strategies and how to coexist in established and emerging cities. Source: Various sources, The Times, UK.

Fig. 1.13: New City: Taobao Village. Shenzhen Bi-City Biennale of Urbanism\Architecture (UABB) 2018. Video by Liam Young and Alexey Marfin. Soundscape by Forest Swords. Commissioned by the Shenzhen Biennale.

Fig. 1.14: Online retailers as Alibaba expand their ecommerce network into the rural areas of China for further growth. Sources: The Wall Street Journal; Gillian Wong and Loretta Chao, August, 2015.

Fig. 1.15: Large Chinese corporations are setting up drone delivery networks in rural areas. Source: Illustration by Jun Cen, Jiayang Fan, The New Yorker, July, 2018.

Fig. 2.01: The poster gives greater visibility to farmers, factory workers and the army. Urbanites are encouraged to embrace a rural work ethic and livelihood strategies. Poster: 'Chairman Mao Zedong Vast World of Accomplishments'. People's Fine Arts Publishing House,1967. Source: Etsy Vintage Reprints.

Fig. 2.02: The diagram shows the Urbanization Rate & Urbanization Growth rate in four phases of Chinese history. Source: Kuangdi, X., 2106, Scipedia, SL, 2018.

Fig. 2.03: 'The people's communes are good ' by Rui Guangting,1958. The artwork for schools illustrates the ideal people's commune, with a healthy harvest, communal facilities for sustenance and sanitation, a centre for the older population and the people's army. Shanghai educational publishing house. Part of the IISH, Stefan R. Landsberger Collections & Anchee Min, Sinological Institute Leiden University.

Fig. 2.04: "Educated youth must go to the countryside to receive reeducation from the Poor and Lower Middle peasants!," 1969. Source: part of the IISH / Stefan R. Landsberger Collections, and Ancee Min.

Fig. 2.05: A map of Greater China with the Coastal cities emphasized as the main population centers driving a sea-based, export-led economy since the late 1970's. A turn to the interior provinces will drive economic growth and urbanization through the development of medium and smaller sized cities. Image sources: by the author, CEIC, WTO.

Fig. 2.06: The Chinese poster presents advancements and innovation in science, technology, space exploration and genetic research. Credits: Author unknown. Publisher: Shanghai renmin meishu chubanshe, Feb. 2001. Part of the IISH / Stefan R. Landsberger Collections, Sinological Institute Leiden University.

Fig. 2.07: National New type Urbanization Plan (20142020): For the first time in 2014, an official national urbanization plan was issued by the Central Government. The policy, initially, for five years seeks to address coordinated regional development. It aims to tackle unintended consequences to human and social issues as a result of urbanization. Furthermore, its objectives are (1) Balanced Urban-Rural development, (2) Coordinated Regional Development, (3) Environmentally friendly and Sustainable Urbanization, (4) Biodiversity Preservation, (5) Food Independence, (6) Poverty Eradication, and (7) Preservation of Ethnic Identities. Credits: Fang, Ma et Wang 2015.

Fig. 2.08: In the next phase of urbanization in China, growth will be mainly in the number of Medium size cities (1.55 million people), Small size cities (0.5 to 1.5 million people), and Big Towns (> 0.5 million people). China counted approx. 858 cities in 2005. By 2025, the number of cities could rise to 939, out of which 8 will be Megacities with over 10 million urban citizen each. Source: McKinsey & Co. report "Preparing for China's urban billion" March 2009.

Fig. 2.09: The diagrams presents some of the known functions of mountains, as new forms of urbanity emerge new ecologies become embedded into the mountain regions. Source: Hans Vejre, M. Antrop, U. Mander.

Fig. 2.10: As coined by IBM's Research Precision Agriculture, 'Cognitive' Precision farming technologies are designed to coordinate the timing for crop harvest informed by weather data and the commodity price volatility. Source: IBM Research Precision Agriculture, Kevin Lopez Alvarez, 2017.

Fig. 2.11: Clean Energy farms in the Chinese countryside. On the interim, due to recent US trade tariffs, the installation of Renewable Energy infrastructure may be scaled-down. Source: Eric Ng, SCMP, March, 2018.

Fig. 2.12: The use of drones for spraying of pesticide in China's Shanxi Province. Source: Various Sources: Xinhua, Cao Yang; News.cn, Global Times, 2017.

Fig. 2.13: Over-consumption of fresh water by wealthier cities of the Greater Bay cluster of cities causing future risks water shortage risks to the economically vulnerable population in the Dongjiang River Basin and the more developed cities in the territory. Additionally, environmental pollution, flood risks, potential future draughts undermine the resilience of the territory for unanticipated natural disasters. Source: SCMP, Ernest Kao,2015.

Fig. 2.14: Mei County in Guangdong province. The maps shows the hydrological system and indicative extent of Hakka and Cantonese ethnic distribution in the province. Image generated by the author, 2019.

Fig. 2.15: Water contamination data in the territory. Image Source: Ding, Jiang, Liua, Hou, Liao, Fud, Peng, 2016.

Fig. 2.16: The water reservoirs are a key ecological resource for the Greater Bay territory. Source: Evergreen, SCMP 2011.

Fig. 2.17: The Fengshuba reservoir was visited during field trip visits. The region can be reached in one hour by car from Meizhou which is one hour flight away from the Guangzhou Megacity. Photography: by the author.

Fig. 2.18: Tea is an important export product and tea plantations as parks frequented by holiday makers.Photography: by the author.

Fig. 2.19: Picture taken in a tea house in the Yannanfei Hakka tea plantations outside of Meizhou, in Mei Xian. Photography: by the author.

Fig. 2.20: Villages and Mountain Temples were visited during field trips guided by a local tour guide. Photography: by the author.

Fig. 2.21: Interior views of courtyard and the living room in visitors centre. Photography: by the author.

Fig. 2.22: Railway Infrastructural Integration in China: A new connection from Hong Kong West Kowloon Station, the Express Rail Link (XRL) is linked with the countrywide high-speed rail network in Greater China. In total the rail network is 25,000km long and constitutes of eight vertical and eight horizontal lines integrating the larger urban centers in China. Sources: Howchou & Express Rail Link (XRL), Transport and Housing Bureau.

Fig. 2.23: Tourism in the Chinese countryside is targeted at Chinese urbanites and visitors from abroad.Source: National Geographic Chinese Edition, 2016, TimeOut Shanghai, 2016.

Fig. 2.24: Thematic categorization of Cultural, farming and culinary tourist experience can be selected on the internet and travel agents. Various sources: TimeOut Shanghai, 2016.

Fig. 2.25: Walled Hakka Villages are popular tourist attractions. Source: Jimmy Ng.

Fig. 2.26: Hakka cuisine restaurant and menu visited during field trips. The restaurant in Chiguang Town specializes in traditional Hakka recipes. Source: by the author.

Fig. 2.25a: Tulous round buildings are an important cultural asset of the Hakka Diaspora in Southeast China. Many of the earth towers were built between the 13th and 20th centuries.Today guided archiectural tours provide access for visitors into the fortified Hakka villages.Tulous can house up to 600 people. Some of the Tulous offer accommodation for tourist to experience the unique Hakka history and culture. Source: Huang et Hart; Ronald Knapp; Travelchinaguide.

Fig. 2.26: Hakka cuisine restaurant and menu visited during field trips. The restaurant in Chiguang Town specializes in traditional Hakka recipes. Source: by the author.

Fig. 2.26a:The Hakka ethnic group from Guangzhou had five major phases of migration in history: Phase 1 317-819; Phase 2 880-1126; Phase 3 1127-1624; Phase 4 1645-1867; and Phase 5 from 1867. Today Hakka people are dispersed all over the world and continue to form distinct immigrant communities outside of China. Source: Lei Hua, Chongcheng, Chen,Hui, Fang, Xiangxiang, Wanga, AsiaSociety.

Fig. 2.26b: An independent-minded ethnic group: Hakka culture emphasizes education and rebellious temperament. Sun Yat Sen was one of the most revolutionary politician who is known as the "the father of modern China". He became the first President of the Republic of China and founder of the Kuomintang (Nationalist Party of China).The above time line shows Sun Yat Sen's life recalling his revolutionary and political struggles. Sources: Repro-tableaux, Hsu Chung-mao for ThinkChina, CHSA Museum. https://chsa.org/exhibits/online-exhibits/sun-yat-sen-an-american-legacy/life-and-times-of-dr-sun-yat-sen/

Fig. 2.26c: Community building: Hakka collective housing examples and ideal communities in the countryside continue to influence settlement planning and housing projects in China.

Fig. 2.26d: Dr. Sun Yat-Sen has left traces in society, politics and culture inside and outside of China. Several monuments commemorating Sun Yat-Sen as the 'father of modern China' are popular tourist destinations attracting visitor from all over the world. Dr. Sun Yat-Sen's legacy continues to be a cultural asset for promoting ethnic-cultural tourism in China.Sources:China Heritage,Lu Yanzhi Architect, Yao Qian and Gu Bing ,Yang Bo China News Service, Portrait photo author unknown,Jiangsu Province Radio and Television Station, Coffin photograph by Lothar Ledderose.

Fig. 2.26e: Chinese ink landscape painting shows the Sun Yat-sen Mausoleum on Zijin Shan - Purple Mountain. Source: Fu Boashi, 1962.

Fig. 2.27: A local interviewee met in Longbei Town during the fieldwork phase of the research. The interviewee was collecting goods at the Taobao distribution centre in her town. Source: by the author.

Fig. 2.28: A liquor shop owner met in Chiguang Town during the interviews with inhabitants. Source: by the author.

Fig. 3.00: The Photograph shows the elevation relief and topographic features of the coastal zone. Source: GoogleEarth, June 2018.

Fig. 3.01: New Synergies of Hong Kong's financial and professional services sectors, Shenzhen's high-tech manufacturing and innovation skills, and the manufacturing strengths of Dongguan and Guangzhou. The development of the area should also act as a catalyst for China's Belt and Road initiative–an ambitious strategy that aims to link the economies along the Silk Road Economic Belt (Central Asia to Europe) and the Maritime Silk Road (South Asia to Africa and the Middle East) together. Source: Fung Business Intelligence, 2016.

Fig. 3.02: The coordinated economic development of the Greater Bay Area is to maximize the competitive strength of all cities in the region. The Pearl River Delta is seen as key region for China's innovation driven development. The areas of collaboration include (1) Innovation and Technology, (2) Financial Services, (3) Transportation and Logistics, (4) Professional Services, (5) International Legal and Dispute Resolution Services, (6) Clearance Facilitation, (7) Medical Services, (8) Education, (9) Art, Culture and Creative Industries, (10) Tourism, (11) Environmental Protection and Sustainable Development, (12) Youth Development. Source: Develop Bureau, Planning Dept. Oct 2016; www.bayarea.gov.hk.

Fig. 3.03: The Pearl River Delta is gradually merging into a single megacity. Source: Samuel Lai, TimeOut HK, 2016.

Fig. 3.04: Overall Territory: Action Plan for the Bay Area of Pearl River Estuary. Source: Hong Kong Planning Department, 2011.

Fig. 3.05: Key City Hallways: Action Plan for the Bay Area of Pearl River Estuary. Source: Hong Kong Planning Department, 2011.

Fig. 3.06: Cross-boundary Transport system: Action Plan for the Bay Area of Pearl River Estuary. Source: Hong Kong Planning Department, 2011.

Fig. 3.07: Green Network: Action Plan for the Bay Area of Pearl River Estuary. Source: Hong Kong Planning Department, 2011.

Fig. 3.08: This illustration shows the water supply reservoirs of the PRD mega city in a consolidated axonometric drawing. At a macro scale, the drawing intends to synthesize the underlying Hakka cultural ethnicity, the hydrological capital and rail network linkages found in the region. Due to the overlapping nature of regional ecological, cultural and infrastructural assets, illustration exemplifies the difficulty of determining definitive borders for the regional features, which transcend geographic boundaries found in ordinance survey maps. The small DEM (digital elevation model) of the reservoirs are enlarged in the following pages for better viewing. Sources: Google Earth & Image generated by the author, 2019.

Fig. 3.09: Digital Elevation Model 01 Pearl River Delta Plain. Refer to Fig. 3.08 for context. Sources: Google Earth & Image generated by the author, 2019.

Fig. 3.10: Digital Elevation Model 02 Lower River Basin at Dongguan. Refer to Fig. 3.08 for context. Sources: Google Earth & Image generated by the author, 2019.

Fig. 3.11: Digital Elevation Model 03 Huizhou Valley Dongjiang River. Refer to Fig. 3.08 for context. Sources: Google Earth & Image generated by the author, 2019.

Fig. 3.12: Digital Elevation Model 04 Xiangang Reservoir Water Supply Valley. Refer to Fig. 3.08 for context. Sources: Google Earth & Image generated by the author, 2019.

Fig. 3.13: Digital Elevation Model 05 Fengshuba Reservoir Water Supply Valley. Refer to Fig. 3.08 for context. Sources: Google Earth & Image generated by the author, 2019.

Fig. 3.27: Elevation map 100 1400m.Source: by the author,Ding, Jiang, Fu, Liu, Peng, Kang, 2015

Fig. 3.28: Urban land use based on administrative boundaries. Map by the author, based on data by Ding, J., Jiang, Y., Fu, L., Liu, Q., Peng, Q., & Kang, M. (2015).Impacts of land use on surface water quality in a subtropical River Basin:a case study of the Dongjiang River Basin, Southeastern China. Water, 7(8), pp. 4427-4445.

Fig. 3.29: Agricultural land use. Map by the author, based on data by Ding, J., Jiang, Y., Fu, L., Liu, Q., Peng, Q., & Kang, M. (2015).Impacts of land use on surface water quality in a subtropical River Basin:a case study of the Dongjiang River Basin, Southeastern China. Water, 7(8), pp. 4427-4445.

Fig. 3.30: Map of soil fertility. Map by the author, based on data by Ding, J., Jiang, Y., Fu, L., Liu, Q., Peng, Q., & Kang, M. (2015). Impacts of land use on surface water quality in a subtropical River Basin:a case study of the Dongjiang River Basin, Southeastern China. Water, 7(8), pp. 4427-4445.

Fig. 3.31: Per capita income level in USD/p.a. Source: Carlos Lo, 2010.

Fig. 3.32: Scenario for an ecology remediation strip along highly polluted river network.. Image Source:by the author.

Fig. 3.33: Eco-remediation zone shown with classification of water quality. Image Source: by the author.

Fig. 3.34: Food security: more agricultural production by increasing the farm land of the existing agricultural operations. Image Source: by the author.

Fig. 3.35: The map suggests a farm land protection zone in-between the mega-city and the mountainous hinterland to maintain independent food production capacities. Image Source: by the author.

Fig. 3.36: The finer grid in red shows distribution of low-density settlement structures. For the sources of the concept of Isotropy in urbanism refer to: Viganò, P. (2008). Water and Asphalt, the Project of Isotropy in the Metropolitan Region of Venice. Architectural Design, 78(1), 34-39. Image Source: by the author.

Fig. 3.37: This scenario maintains the megacity region, while securing a consolidated farm land protection zone and dispersed urbanization in the mountains. Image: by the author.

Fig. 3.38: While maintaining the existing agricultural ecologies, this scenario proposed a horizontal distribution of economic ecologies which include both rural and urban livelihood activities. Image: by the author.

Fig. 3.39: The scenario proposes the insertion of small cities in the economically disadvantaged territories. It suggests the resettlement of a displaced migrant population from larger cities into new communities. The setting up of new communities would require support from the government and the need to attract investment capital. Large-scale agricultural companies in partnership with Provincial and Township level government could collaborate in development associations to attract the necessary funds for urbanization.

Fig. 3.40: The hypothetical scenario suggests that population will relocate from overcrowded city centers to smaller cities. New cities will be eco-friendly, people-oriented and have independent food production capacities. Image Source: by the author.

Fig. 3.41: Nestling of new settlements based on the topography. Image Source: by the author.

Fig. 3.42: Light red mesh shows dispersed communities in the flat valley of the topography. In this scenario the steeper mountain topographies are preserved for forestry. Image Source: by the author.

Fig. 3.43: Green outlines highlight landscape and scenic mountain regions with potential as tourist destinations. Farming communities are nestled in the valley regions in-between the mountains. Image Source: by the author.

Fig. 3.44: Three main sections through the territories show the relationships from the Megacity to the mountains in the hinterland. Transect: by the author.

Fig. 3.45a: Part 1 of Table. Categorization of anthropogenic activities and potential new ecologies for the study sections. See potential interventions for a transformation shown in

Fig. 3.46 - .3.48. Sources: Griboedov and by the author.

Fig. 3.45b: Part 2 of Table, see Fig. 3.45b. For the preparation of the potential interventions the objectives of the UN Sustainable Development Goals (SDG) 2015-2030 have been taken into consideration, see below summary list of SDG for reference.

Fig. 3.46: The Upper section shows the existing ecologies the below sections proposes the embedding of new ecologies for greater independence and environmental remediation of the territory. Transect: by the author, pictogram: based on Griboedov.

Fig. 3.47 & Fig. 3.48: refer to caption in Fig. 3.46. Transect: by the author, pictogram: based on Griboedov.

Fig. 3.49: The diagram illustrates a condition of Dependence for communities with its causes, associated negative externalities, consequences, social and environmental damages. Diagram: By the author.

Fig. 3.50: Also see Fig. 3.85. Three study sites along a transect, Chiguang town close to Fengshuba reservoir, Longbei town at Heshui reservoir and Changsha town in Mei county. 3D-printed stereo-lithography of terrain: by the author.

Fig. 3.51: Chiguangzhen town context with Fengshuba Reservoir: Sections through the topography overlaid on 3D-printed terrain: The sections in 10km intervals indicate the rivers and water reservoirs,with the map showing where settlements are located. Further, the sections identify the slopes < 5% (potentially suitable for farming activity, provided fertile soil is in place on site), and slopes > 10% are shown (not suitable for farming, unless terracing of topography is considered). Images: by the author.

Fig. 3.52: Map of Chiguangzhen town larger context with hydrological system, main roads and settlements shown. Sources: Google Earth & Image by the author with satellite map underlay.

Fig. 3.53: The 3D visualizations show the settlement morphology of Chiguangzhen. The settlement is organized along a main road with in-line shops. Sources: Google Earth & Images by the author.

Fig. 3.54: Photographs F01 - F09: Fengshuba Reservoir dam and surrounding farms. Images: by the author.

Fig. 3.55: Chiguangzhen, Photograph F10: Hakka courtyard houses are over one hundred years old. The spaces are open for community gatherings and for tourist visitors. Images: by the author.

Fig. 3.56: Chiguangzhen Town, Photographs F11 - F19: Images: by the author.

Fig. 3.57: Chiguangzhen, Photograph F20: Traditional Hakka courtyard house. The building is listed as a heritage building. Images: by the author.

Fig. 3.58: Chiguangzhen Town, Photographs F21 – F29: Images: by the author.

Fig. 3.59: Chiguangzhen, Photograph F30: Although Hakka courtyard houses are open to visitors, the buildings are still inhabited by local families. Image: by the author.

Fig. 3.60: Land use Map of Chiguangzhen town with built up area, agricultural land, roads, rivers, lakes and forestry shown. Sources: Google Earth & Image by the author with satellite map underlay.

Fig. 3.61: Chiguangzhen Town with Fengshuba Reservoir is shown in axonometric section. The current livelihood activities are a mixture agricultural and construction material production. Illegal Rare Earth extractions of Dysprosium, Terbium and Europium are causing damage to agricultural land due to the acid washing of the soil. Ground water contamination is also an issue in this settlement. Rare earth extractions are sold to refineries for the production of electronic goods. Talking to local inhabitants, they noted the community needs better facilities for disabled people. Currently, abandoned shops are used as schools for children with special needs. Sources: Google Earth & Images by the author.

Fig. 3.62: Longbei town context with Heshui reservoir, and Xingning's dispersed settlement structure: Sections through the topography overlaid on 3D-printed stereo-lithography of terrain: Images: by the author.

Fig. 3.63: Map of Longbei town larger context with hydrological system, main roads and settlements shown. Sources: Google Earth & Image by the author with satellite map underlay.

Fig. 3.64: The 3D visualizations show the settlement morphology of Longbei. The town has in-line shop front streets connected with secondary back streets. Sources: Google Earth & Images by the author.

Fig. 3.65: Longbei town, Photographs X01 – X09: Images: by the author.

Fig. 3.66: Longbei, Photograph X10: Longbei's wet markets require upgrading. Several of the shops are not in use or have inadequate sanitary facilities and electricity provisions. Images: by the author.

Fig. 3.67: Longbei town, Photographs X11 – X19: Images: by the author.

Fig. 3.68: Longbei, Photograph X20: Local vegetable shop attendant loading orders into a small van for delivery. Shops in the main retail street are in deplorable condition. Images: by the author.

Fig. 3.69: Longbei town, Photographs X21 – X29: Images: by the author.

Fig. 3.70: Longbei, Photograph X30: View of Longbei's streetscape. Most of the buildings are two story-high buildings. Taller town houses are up to four stories high. Images: by the author.

Fig. 3.71: Land use Map of Longbei town with built up area, agricultural land, roads, rivers, lakes and forestry shown. Sources: Google Earth & Image by the author with satellite map underlay.

Fig. 3.72: Longbei Town with Heshui Reservoir is shown in axonometric section. In this town acute poverty and the pollution can be witnessed when walking through the main streets and visiting the shops. This settlement also has a basic Taobao distribution shop, which is part of the Alibaba poverty eradication program for disadvantaged communities. There was only one employee and two customers when visiting the Taobao collection shop. The local wet market was practically empty during the visit, and in need of urgent upgrading. Illegal mining for rare earth is also practiced. The Heshui reservoir could be promoted as a tourist destination. However, raising external sources funds for setting up tourism facilities such as a water sports clubs and camping sites would need to the considered. The provincial government, Chinese corporations, charities or private donors could be approached for subventions. Sources: Google Earth & Images by the author.

Fig. 3.73: Changshazhen town context with Meijiang River flowing through Meizhou: Sections through the topography overlaid on 3D-printed stereo-lithography of terrain: Images: by the author.

Fig. 3.74: Map of Changsha town larger context with hydrological system, main roads and settlements shown. Sources: Google Earth & Image by the author and satellite map underlay.

Fig. 3.75: The 3D visualizations show the settlement morphology of Changshazhen along the Meijiang River. The mixed agricultural and industrial activities characterize this settlement. Small scale family farms typically have three-to-four story dwelling buildings for the older and younger generation of the family. The farming fields are surrounded by industrial sheds, construction material storage sites, new family houses and derelict traditional farms houses. Sources: Google Earth & Images by the author.

Fig. 3.76: Changshazhen town, Photographs M01 – M09: Images: by the author.

Fig. 3.77: Changshazhen, Photograph M10: View along Meijiang River with semi-completed dwelling units and industrial sheds. Images: by the author.

Fig. 3.78: Changshazhen town, Photographs M11 – M19: Images: by the author.

Fig. 3.79: Changshazhen, Photograph M20: View of scattered farm buildings to forests in Changshazhen. Images: by the author.

Fig. 3.80: Changshazhen town, Photographs M21 – M29: Images: by the author.

Fig. 3.81: Changshazhen, Photograph M30: Traditional single-to-two story Hakka farm houses can be found in nestled in the landscape in-between rice fields and newer family houses. Images: by the author.

Fig. 3.82: Land use Map of Changshazhen town with built up area, agricultural land, roads, rivers, lakes and forestry shown. Sources: Google Earth & Image by the author with satellite map underlay.

Fig. 3.83: Changshazhen in the Meijiang River Valley is shown in axonometric section. The small family farms and constructional material industrial sheds can be found along the river. The production plants specialize in making bricks and cement-based masonry blocks. Material storage set ups can be seen and delivery trucks area loaded with bricks for dispatch. The low-density settlement has some advantages, for example its proximity to Meizhou and the surrounding mountain scenery. The family farms could diversify into agricultural tourism. Few restaurants offering Hakka cuisine can be found. Access and connectivity for delivery of good by several Chinese e-commerce companies such as Alibaba and JD is set up. Sources: Google Earth & Images: by the author.

Fig. 3.84: The Yannafei Tea plantations in Mei country is a popular tourist destination attracting visitors mainly from Guangdong as well as neighbouring provinces in China. The tea plantations were visited as part of the fieldwork trips in the region. Talking to a local mountain dweller illegal digging of wells for water and extractions of metals is an issue for local farmers. Sources: Google Earth & Images: by the author.

Fig. 3.85: Also see Fig. 3.50. Five Livelihood Scenarios of interviewees: Shown along a section through the Mei county territory in the context of Chiguangzhen, Longbeizhen , Changshazhen study sites. Sources: Google Earth & Images: by the author.

Fig. 3.86: Formal and Informal Interviews with local residents were conducted through a translator. The local inhabitants speak the Hakka dialect, linguistically related to Cantonese. Photography: By the author.

Fig. 3.87: 'Independence' is the ideal condition for remote settlements and the inhabitants. The diagrams captures the key determinants and qualities needed for independent livelihoods.

Fig. 3.88: While ideally, the five to six of the independence objectives would be met for a relatively balanced livelihood, at most the five local characters met 13 of the determinants. Therefore they have been grouped in higher Dependence Levels of 3 – 5. Source: by the author.

Fig. 3.89: Yu is a small canteen owner offering meals for local residents. He has enhanced his business by specializing in local signature meals. He sources all the produce through a local network of organic farmers. Additionally, he runs a delivery service for cooked meals and organic farming produce. Source: By the author.

Fig. 3.90: Huiru set up a clean energy operated, off-grid Eco-lodge. She is also a smart farming consultant and Sustainability school teacher. Source: By the author.

Fig. 3.91: Han is a guest house and tea house owner. He is also involved in community services. Additionally, he is a driver for electric mini buses service for elderly and has a drone pilot license. Source: By the author.

Fig. 3.92: Yang is a tour guide in a mountain temple complex. He grows and sells Chinese medicine produce. Occasionally, he also works as a contract farmer on local farming collectives. Source: By the author.

Fig. 3.93: Fei is a Tourist Bus driver to local Agricultural Tourism farms. He also guides visitors to local water reservoir recreational sites. He also sells and delivers farming produce from the farms he visits. Source: By the author.

Fig. 3.94: Shop front of a Rural Taobao service center in Yufeng village (left). Taobao delivery van sourcing agricultural produce from farmers in Yufeng village for selling goods through Alibaba's online retailing platform (right). Image Source: Julie Makinen/ L.A. Times, 2016.

Fig. 3.95a: Alibaba Poverty Relief Fund announced by Alibaba's CEO Jack Ma in 2017 set up 10 billion yuan for poverty alleviation in rural and marginal mountain regions. Image Source: Xinhua 2017 / www.news.cn.

Fig. 3.95b: Distribution of Taobao Villages in China, 2014-2018. In 2019 Alibaba reported over 4000 Taobao Villages have been created. The maps by AliReseach show most of the e-commerce distribution centres in provinces along the coastal region of China. There are very few Taoboa villages set up in the poverty-stricken, land-locked Central and Western provinces of China. Image Source: AliResearch, 2018.

Fig. 3.95c: A Taobao distribution centre was visited during one of the fieldwork trips to Longbei town close to Heshui Reservoir. The shop is equipped with only one PC workstation and flat bed monitor, and a small counter for receiving customers. The set up of the shop with shelves for delivery goods is basic.Sources: Lulu, F.(2019),Taobao Villages - The Emergence of a New Pattern of Rural Ecommerce in China and its Social Implications, Retrieved 16.09.2019,https://www.fes-indonesia.org/e/en-taobao-villages-the-emergence-of-a-new-pattern-of-rural-ecommerce-in-china-and-its-social-implications/; Axonometric and photograph generated by the author.

Fig. 3.95d: The same e-commerce giants that claim to commit to poverty reduction, provide the on-line platforms for selling cheap smart phones with touch screens produced with rare earth extractions from impoverished rural and mountain areas. Extraction of rare earth for electronic products are often mined by criminal gangs in China, entering farms and illegal digging out metals and acid washing soil intended for agricultural cultivation.Image Source: Handsetrecycle.

Fig. 3.96a: As a further development of the New Type Urbanization Plan beyond 2020, the following focus areas have been identified for the research by design studies for the Present to 2050+. The development objectives have selected for the chosen study sites to best achieve emancipation and to become more economically and resource independent settlements: (1) Renewable Energy Independence, (2) Pollution Reduction, (3) Food Independence, (4) Access to Information and Education, (5) Economic Diversification, (6) Social Security. Sources: Xu, Wang, Zhou, Wang, Liu, 2016; and by the author.

Fig. 3.96b: The timeline diagram for the present to 2050+ proposes incremental local interventions for the making of local ecologies and an emancipated urbanity. Neither the sequence nor the types of interventions are prescriptive. The main purpose for planning for the next phases of urbanization is to have repertoires of possibilities prepared in times of uncertainly. Source: By the author.

Fig. 3.96c: Principle of Alternative possibilities. The table lists potential interventions in distressed territories. Here table is provided for purpose of summary, for larger view see Fig. 3.45 a & b.

Fig. 3.96d: Giving support to marginalized communities on a path to sustainable development entails the provision social and welfare services which are part of a planning process. In particular, it considers vulnerable members of the community including please with health issues, children, mothers, disabled people and the elderly. Clean energy and energy independence is a shared responsibility for settlements in a process of adaptation for climate change.

Fig. 3.96e: This table of livelihood determinants shows inclusion of recent technologies and know-how as part of an innovation and diversification strategy. The strategies include Informatization of remote territories, Drone mobility systems, opening of new distribution channels, alternative clean energy, Smart farming practices, Eco-remedial technologies, E-commerce, and Social media networks. Image Source: by the author, adapted based on Ellis, 2000.

Fig. 3.96f: The multiplication, networking and exchanges of new livelihood activities will overtime grow emancipated local ecologies. Source: By the author.

Fig. 3.96g: While territories must sustain systemic independence to a certain degree, interrelations with different urban agglomerations are the feature of an internationalized economy. For the maintenance of the long-term stability for both megacity agglomerations and dispersed mountain communities Interdependence between territories is an unavoidable necessity. The strategic preconditions for the coexistence of different types of urbanities in a globalized world are shown. Diagram: By the author.

Fig. 3.97a: Scenario construction Changsha town: The case of Mr. F's livelihood strategy is presented in the context of Changshazhen. Fei is a guide for visitors to agricultural-tourism and mountain scenic sites. Additionally, he works as a driver and is a drone operator for a delivery service. In the case of Mr. F's agri-tourism farms smart technologies have been embedded as shown in the section to support a future-oriented transformation of the settlement. Sources: Google Earth & Images by the author.

Fig. 3.97b: Scenario construction Longbei town: The case of Mr. L's livelihood strategy is taken as an example in the context of Longbei. Mr. L. specializes in Hakka cooking and distribution of farming produce. As shown in the section axonometric, his town has undergone a process of transformation by embedding several reforestation projects to restore a fragmented habitat, eco-farm clusters for sustainable farming, promotion of Heshui reservoir as a tourism destinations, the provision of schools and better welfare services for the elderly, due to its aging population.Sources: Google Earth & Images by the author.

Fig. 3.97c: Scenario construction Chiguang town: The case of Mrs. L's livelihood strategy is shown an example in Chiguangzhen. Mrs. L. is a teacher by training and is interested in sustainability education. Besides teaching in the local school, she also offers advice to farmers on renewable sources of energy and water decontamination. The development strategy for Chiguangzhen is based on providing better education facilities, teaching sustainability practice for farming and energy generation in primary and secondary schools. The community also set up a clinic and nursing home to provide access to basic medical care for all members in this settlement. Sources: Google Earth & Images by the author.

Fig. 3.98a: The revitalization of Chiguang town's main street proposes the overhauling of shop fronts, greening of roofs and installation of street lighting for improving safety at night-time and to enable the use of outdoor public spaces during evening hours for business activities and recreation. Additionally, the planting of trees, installation of vertical vegetation systems to building facades, the creation of new small parks in the back streets and underutilized plots is proposed. These interventions are intended to increase the vegetation coverage in the town. Gradual adaptation to re-newable energy generation is planned for by the incremental installation of wind turbines for the community. Farming plots are consolidated into larger fields to increase the efficiency of farming activity and secure independent food production for the town. Image: By the author.

Fig. 3.98b: The section shows the overhauled ground floor shop fronts with new awnings for sun shading. New Street lighting for safety at night time is suggested. The paving zones are widened to allow for people to do business outdoors. The building facades have vertical greening and green roofs are proposed to support biodiversity in the town. Image: By the author.

Fig. 3.98c: This example of transformation illustrates the conversion of an abandoned ground floor shop into a Community Walk-in clinic. The idea is to provide basic medical care to members of the community. To attract the local inhabitants and to promote public health, the clinic is located on the main road of this settlement facing a community a market place for easy access. The newly created market place has tree planting to create a pleasant gathering place. On the plaza, market stalls can be installed for selling local farming produce and food which people can consume at outdoor canteens. The roof areas of the buildings are utilized for the installation of solar panels and green roofs systems. Image: By the author.

Fig. 3.98d: For the transformation of Longbei town, new farming parks for agricultural tourism are shown in this example.

Fig. 3.98e: The narrow alley ways are transformed into upgraded shopping streets by planting trees, installing street lighting and awnings at shop fronts. The smaller roof tops are converted into gardens for inhabitants living on the upper levels of the small live and work buildings. It is suggested that some farming activities for personal consumption could be carried out on roof the gardens in addition to running a shop on the ground floor. Image: By the author.

Fig. 3.98f: In this example two interventions for the community are proposed, which is the creation of a new indoor wet market, a public sports facility and a physiotherapy center for local inhabitants. Abandoned small buildings are combined for the reuse as larger indoor spaces for the sports facility. Here people can gather for team sports and exercise classes. Additionally, physiotherapists have training and consultation rooms.

The upper floors facility is intended to improve the well-being and health of the inhabitants of this town. On the ground floor, the joined-together shops create new linkages between existing streets and an indoor market place with a food hall. Image: By the author.

Fig. 3.98g: The mixed-use industrial and farming community of Changsha town is organized in streets running parallel to the Meijiang River. However, the settlement has its main shop front streets facing away from the river front. Additionally, the build-up area has limited linkages to the farm land behind the buildings of this town. The design proposal creates green parks, green roof areas and the planting of new trees in-between the existing buildings. This is to connect the cultivated farmlands to the revitalized shopping streets with intermediate open green park areas as much as possible. In other words, the idea is to create a better mixed transition between the built-up zone and the farming plots by introducing new parks in-between. The shops of the two-three story buildings are redecorated and renovated to be attractive to tourist and the local community. Further, new smaller parks along the river bank are created to draw people and recreational activities to the river front. Image: By the author.

Fig. 3.98h: This street transformation section shows the creation of a "Green Lung" in-between the buildings on an undeveloped plot of land. The newly planted trees contribute the absorption of pollutants in the air and the cooling of temperature in the summer by providing shading. The shops are refurbished and a 7m paving zone between building front and the new park allows for restaurants and coffee shops to offer outdoor seating areas looking out at the new tree planted park. Vertical green walls are suggested for the existing buildings and the roof areas are converted into gardens. Image: By the author.

Fig. 3.98i: The river bank site has been transformed into a park facing a former warehouse retrofitted into a mixed use building with a kindergarten on the ground floor, a Mother & Child care center for families in need on the first floor. The upper floors have been converted into low income housing flats. The building is partly serviced by renewable energy from new solar panels installed on the roof. Opportunities for creating roof top green areas are maximized. A 7m wide street provides access from the kindergarten to a three-story building dedicated to a small community center with a library and computer laboratory. The community center gives people free access to the internet and computer literacy training courses. Image: By the author.

Fig. 4.01: Diagram of a cross-section of the earth's crust. Alexander von Humboldt was a German 19th Century Enlightenment thinker, explorer, geographer and polymath, who used large-scale geographic sections to explain and map geological information and ecological knowledge. Source: Alexander von Humboldt, 1841.

Fig. 4.02: In the East Asia context, Desakota has been coined by Terry McGee, which describes an extended metropolitan region of interlinked networks of rural-urban transactional spaces. The illustration to the right by Terry McGee shows a transactional space in a globalized economy. The diagram to the left shows a spatial arrangement of hybrid and mixed activities in a mega urban region as described by McGee, 2008.

Fig. 4.03: Geddes triad of WORK, PLACE and FOLK from Cities in Evolution: an introduction to the town planning movement and the study of civics, Williams & Norgate, 1915. The illustration to the right shows a triad of Ecology, Livelihood and Community is an assimilation for this study by the author.

Fig. 4.04: A transect by Ian McHarg from his book Design with Nature published in1969. As a landscape architect McHarg advocated for ecological planning as an intrinsic part of regional planning. The transect is used to coordinate ecological planning and land-use planning n multiple overlapping ecological systems: Ian McHarg, 1969.

Fig. 4.05: L'organisation de l'espace by Jean Labasse, and It territorio dell' architettura by Vittorio Gregotti advocate the expansion of the scope of the architectural discipline to design in larger scale territories. The origins of this hypothesis can be traced back to post WWII European national reconstruction efforts of economies. Given the current Global Challenges including climate change, food insecurity, poverty and environmental crisis research projects in larger scale territories are recognized as anurgency for designers.

Fig. 4.05a: The Three Magnets diagram: Town, Country and Town-Country and 'Group of Slumless, Smokeless Cities'diagram. Sources: "To-morrow: A Peaceful Path to Real Reform" - Ebenezer Howard,1898 and the Int.Garden City Institute.

Fig. 4.05b: Welwyn Garden City in Hertfordshire by Howard, an example for better community planning. Source:"To-morrow: A Peaceful Path to Real Reform" - Ebenezer Howard,1898.

Fig. 4.05c: The diagram for a garden city shows 1000 acres of town land and 5000 acres of agricultural land. (6,000 acres or 2,400 ha). A garden city would be planned for 32,000 people. A cluster of six satellite garden cities surrounding a town center of 50000 people would form a group of garden cities see Figure 4.05a. Letchworth, the first garden city was built in 1899 by the Garden Cities Association founded by Howard. The objective was to provide better living conditions for the working class living in 'crowded, unhealthy cities'. Source: "To-morrow: A Peaceful Path to Real Reform" - Ebenezer Howard,1898.

Fig. 4.06: The SmartCode developed by Andres Duany, Sandy Sorlien, William Wright uses the urban-to-rural transect as a planning tool and proposes a planning code based on structured transect zones. Although a highly idealized approach to a gradient City – to - Countryside planning tool, the systematic approach can to a certain degree assist as a framework for working in highly fragmented and hybrid urban-rural territories. Source: by Andres Duany, Sandy Sorlien, William Wright, 2008.

Fig. 4.07: The Valley Section by Geddes in 1909, explores the merging of rural and urban. It discusses the various settlement types, extractions of resources, occupations of inhabitants in their physical environments of the region and how they interact. Geddes believes the " town arises and renews itself from the country...in character, individual and social.". Source: Patrick Geddes,1915. Opportunities for new livelihood strategies enhanced by new mobility systems, the informatization of regions, modern agronomy or precision farming, renewable sources of energy and e-commerce platforms have been added to the diagram. This is to suggest the embedding of new know-how and technologies in transects for the transformation of fragmented environments.

Fig. 4.08: 'A Geddes Valley Section variant' elaborated with Prof. David Grahame Shane. This stepping valley diagram explains the structure of the hydrological system of the Donjiang River Basin in its larger regional context. The diagram is a way to organize and analyze research work of the different ascending environments and ecologies in the river basin. Due to the large scale of the basin territory and its dependent ecologies it is difficult to capture and describe all of the different environments found in this larger scale stepping valley section. In this research project I chose to concentrate on the remote upstream region as it is poorer than the lower zones of basin closer to the mega-city valley. Image generated by the author, 2019.

Fig. 4.08a: The 86.7 hectares master plan for in Xinyang City, Henan Province, China was conceived in 2020 for the Xinyang University. As a starting point for the masterplan design the topograpgy and hydrological network is the basis for this scheme.Source: Sasaki, 2020.

Fig. 4.08b: The hilly terrain, valley spaces, mixed forests and post-agricultural landscapes have been incorporated into the design.Source: Sasaki, 2020.

Fig. 4.08c: The valleys of the landscape are the spatial and organizing feature accommodating the social amenities of the campus masterplan. Source: Sasaki, 2020.

Fig. 4.08d: Preserving the eco-system services and designing with nature has been part of the planning process.Source: Sasaki, 2020.

Fig. 4.08e: The natural site drainage has been considered to avoid disrupting the existing hydrological system. Source: Sasaki, 2020.

Fig. 4.08f: Several ecological design strategies have been deployed including clean energy generation, geothermal, rain water harvesting, grey water recycling, compost management, natural light and ventilation, passive heating and cooling strategies, EV charging stations, green roofs and permeable paving.Source: Sasaki, 2020.

Fig. 4.08g: To manage the environmental impact of the development and construction activity, the masterplan development has been phased into three stages to preserve the natural areas, trees, and riparian corridor erosion protection has been planned as part of the phasing considerations.Source: Sasaki, 2020.

Fig. 4.08h: The aim of this 843 hectares urban green infrastructure project in Shanghai is to restore the ecology in the urban area, improve its micro-climate and to re-connect green amenities with the urban fabric. Source: Sasaki, 2015.

Fig. 4.08i: The three main ideas guiding the development of the urban park are the restoration of ecology, enhancement of the microclimate and reconnecting the part to the urban fabric. Source: Sasaki, 2015.

Fig. 4.08j: Existing hard edge canals with no biological value are restored into riparian ecology corridors. The remediated riparian corridors form habitats for bio-diversity, and purify surface and groundwater on site. Source: Sasaki, 2015.

Fig. 4.08k: The following Wetland zones have been considered for restructuring the riparian edges: Evergreen broad leaf zone, Emergent wetlands, Shrub wetland, Emergent - Submerged wetlands zone, Near-shore littoral zone, open water limnetic zone. The complexity and diversification of the wetland zone contributes to eco-system resilience. The eco-system resilience also supports biodiversity, which then cultivates a diversified food web for a self-supporting, self-sufficient ecology. Source: Sasaki, 2015.

Fig. 4.08l: The new park is also conceived to regulate the urban micro-climate. This is an adaptation strategy for future global warming and the mitigation of the local urban heat island effect. It is expected that a temperature decrease of 3-4 degree Celsius within 2km of large parks can be achieved. Source: Sasaki, 2015.

Fig. 4.08m: Air purification and pollutant absorption is one of the eco-system services the park will provide to the urban area and communities. The parks woodlands have the capacity to release 1080 tons of Oxygen. The park will absorb 6 tons of PM2.5/PM10, 6.1 tons of Sulphur Monoxide (SO), 6.1 tons of Nitric Oxide (NO), 6.9 tons of Trioxygen (Ozone).Source: Sasaki, 2015.

Fig. 4.08n: Lingang's 546-hectare Green Ring Park - Lingang New City Master plan. Source:Sasaki, 2015.

Fig. 4.08o: The radial organization of the masterplan synthesizes urban development, agricultural land use, green infrastructure, tourism and recreational destinations. The development is intended for 800,000 future inhabitants. Source:Sasaki, 2015.

Fig. 4.08p: South Taihu Lake master plan is a 1,312 hectares planning and urban design project in Huzhou, China.Source:Sasaki, 2014.

Fig. 4.08q: The masterplan integrates traditional villages houses and water sports amenities as part of the productive infrastructure to attract tourism to the local economic activities.Source: Sasaki, 2014.

Fig. 4.08r: Agri-cultural tourism parks and food security for the local population are combined with green infrastructure to maintain the local eco-system stability. Source: Sasaki, 2014.

Fig. 4.08s: The eco-corridor project was initiated for the Ningbo Planning Bureau – East New Town Development Committee in 2014. Source: SWA Group, 2014.

Fig. 4.08t: The 'living filter' – eco-corridor project utilizes the topography and ridge lines of the site to purify, direct water flow and collect rain water. The landscape grading strategies and vegetation planting are used for water cleaning and habitat creation.Source: SWA Group, 2014.

Fig. 4.08u: The riparian wetland zones are structured into cascading ponds collecting and filtering surface and rain water.Source: SWA Group, 2014.

Fig. 4.08v: The following design interventions have been incorporated into the design for the purpose of storm water management: Grassland and woodland areas, bio swale zones, permeable paving areas, rain garden water collection, and surface water collection strategies, water run-off filters and treatment details. The landscaping design includes flood storage capacities and the channelling of clean water back to the ground and stream.Source: SWA Group, 2014.

Fig. 4.08w: The completed eco-corridor project not only serves the local communities in the city but also raises awareness of the environmental green infrastructure needed and is an example to follow for future urban regeneration projects in China.Source: SWA Group, 2014.

Fig. 4.08x: Jianjiang Valley, Pengzhou. Project area 150 km^2 Source:Agence Ter, 2018.

Fig. 4.08y: The Jianjiang Valley masterplan is a study of creating synergies between 6 main villages and 3 smaller settlements. One of the objectives was to densify the network of villages, while preserving a balance of urbanization and maintaining rural areas and its ecology.Source: Agence Ter, 2018.

Fig. 4.08z: The urbanization plan in the mountain valley considers the protection of fertile soil for agricultural activity, and carefully manages the transitions between natural ecology to urban areas, while maximizing opportunities for renewable energy generation.Source: Agence Ter, 2018.

Bibliography:

[A]:

Afridi, F., Li, S. X., & Ren, Y. (2015). Social identity and inequality: The impact of China's hukou system. Journal of Public Economics, 123, 17-29.

Afshar, F., (1998). Balancing global city with global village. Habitat International, 22(4), pp.375-387.

Andresosso-O'Callagham, B. and Wei, X., (2003). EU FDI in China: Locational determinants and its role in China's hin-terland. In Proceedings of the 15th Annual Conference of the Association for Chinese Economics Studies, Austra (ACESA).

Anthony, G. et al. (1990), 'Changes in City Size and Regional Distribution, 1953-1986' in Chinese Urban Reform, What Model Now?, edited by Grant Blank et al. London: Sharpe, Inc.

[B]:

Balk, H.H., (1945). Rurbanization of Worcester's Environs. Economic Geography, pp.104-116.

Barabasi, A.L., (2002). Linked: How everything is connected to everything else and what it means. Plume Editors.

Batty, M., Besussi, E., Maat, K., & Harts, J. J. (2004). Representing multifunctional cities: density and diversity in space and time. Built Environment (1978-), 324-337.

Bezzadfar, M., Alalhesabi, M. and Amirhodaei, E., (2017). Typological Analysis of The Transect and Its Background Theories and Approaches.

Belk, R., (2014). You are what you can access: Sharing and collaborative consumption online. Journal of Business Re-search, 67(8), pp.1595-1600.

Benevolo, L. (1967) Nineteenth-Century Utopias, from The Origins of Modern Town Planning. Op.Cit., pp. 39-75.

Berrizbeitia, A. (2015). The Countryside I: Ruralism. Harvard GSD.

Berry, C., et al. (2010), The New Chinese Documentary Film Movement for the Public Record. Hong Kong: Hong Kong University Press.

Bertoni, G., (2015). World food production. La politica.

Bessière, J., (1998). Local development and heritage: traditional food and cuisine as tourist attractions in rural areas. Sociologia ruralis, 38(1), pp.21-34.

Bhaumik, T.K. (2009), Old China's New Economy: The Conquest by a Billion Paupers. London:SAGE.

Bianchetti, C., (2015). Territories in Crisis. Architecture and Urbanism Facing Changes in Europe. Berlino, Jovis.

Bohl, C.C. and Plater-Zyberk, E., (2006). Building Community across the Rural-to-Urban Transect [The Transect]. Places, 18(1).

Brandt, J., & Vejre, H. (2004). Multifunctional landscapes - motives, concepts and perceptions. In J. Brandt, & H.Vejre (Eds.), Multifunctional Landscapes: Volume 1 Theory, Values and History. (pp. 3-32). Southhampton: WIT Press. (Ad-vances in Ecological Sciences, Vol. 1).

Branigan, T. (2014), China reforms hukou system to improve migrant workers' rights,[Online], available from: http://www.theguardian.com/world/2014/jul/31/china-reform-hukou-migrant-workers, (accessed 16.11.15).

Brenner, N. (2000). The urban question: reflections on Henri Lefebvre, urban theory and the politics of scale. Interna-tional journal of urban and regional research, 24(2), 361-378.

Brenner, N. and Keil, R. (eds.) (2006), The Global Cities Reader. New York: Routledge.

Brenner, N. (2013). Theses on Urbanization, Public culture, Vol. 25, No.1, pp. 85-114.

Brenner, N., & Schmid, C. (2014). The 'urban age'in question. International Journal of Urban and Regional Research, 38(3), pp.731-755.

Brenner, N., & Schmid, C. (2015). Towards a new epistemology of the urban?,City, 19 (2-3), pp. 151-182.

Broudehoux, A. (2002), The Making and Selling of Post-Mao Beijing. London: Routledge.

Brunfaut, V. (2002). Towards an Urban for the "dispersed city ". Comunicació presentada a la conferència d'EURA: Urban and Spatial European Policies. Torí, pp. 18-20.

Buchanan, R. (1992). Wicked problems in design thinking. Design issues, pp. 5-21.

Bunnell, T. G., & Coe, N. M. (2001). Spaces and scales of innovation. Progress in Human geography, 25(4), pp. 569-589.

Bunnell, T. (2004), Malaysia, Modernity and the Multimedia Super Corridor: A critical geography of intelligent land-scapes. London: Routledge.

Burdett, D., & Nowak, W. (2011), Well-being in the Urban Age, [Online], available from: https://lsecities.net/media/objects/articles/well-being-in-the-urban-age/en-gb/, (accessed 16.11.15).

Burdett, R., & Sudjic, D. (2007). The endless city: an authoritative and visually rich survey of the contemporary city. Phaidon Press.

Byé, P. and Fonte, M., (1993). New Functions for Rural Space in Western Europe: a Challenge for Agricultural Tech-niques. International Journal of Sociology of Agriculture and Food, 3, pp.81-95.

[C]:

Cai, F. (2011). Hukou system reform and unification of rural–urban social welfare. China & World Economy, 19(3), 33-48.

Calafati, A. G. (2005). From territory to city: the conceptualisation of space in Italy since 1950. Università Politecnica delle Marche, Dipartimento di economia.

Camillus, J. C. (2008). Strategy as a wicked problem. Harvard business review, 86(5), 98.

Campanella, T. (2008), The Concrete Dragon: China's Urban Revolution and What it Means for the World. New York: Princeton Architectural Press.

Canizaro, V.B. (ed.) (2007), Architectural Regionalism. Collected Writings on Place, Identity, Modernity, and Tradition. New York: Princeton Architectural Press.

Cartier, C. (2002), 'Transnational Urbanism in the Reform-era Chinese City: Landscapes from Shenzhen', Urban Studies 39/9: 1513-32.

Chan, K. W. (2010). Fundamentals of China's urbanization and policy. China Review, 63-93.

Chan, K. W. (2010). The household registration system and migrant labor in China: notes on a debate. Population and Development Review, 357-364.

Chan, K. W. (2009). The Chinese hukou system at 50. Eurasian geography and economics, 50(2), 197-221.

Chan, K. W., & Zhang, L. (1999). The hukou system and rural-urban migration in China: Processes and changes. The China Quarterly, 160, 818-855.

Chang, G. H., & Brada, J. C. (2006). The paradox of China's growing under-urbanization. Economic Systems, 30(1), 24-40.

Chang S., and R. Yin-Wang Kwok (1990), 'The Urbanization of Rural China' in Chinese Urban Reform, What Model Now?, edited by Grant Blank et al. London: Sharpe, Inc.

Chen, A., (2002). Urbanization and disparities in China: challenges of growth and development. China Economic Re-view, 13(4), pp.407-411.

Chen, J., Guo, F., & Wu, Y. (2011). One decade of urban housing reform in China: urban housing price dynamics and the role of migration and urbanization, 1995–2005. Habitat International, 35(1), 1-8.

Chen, L. and Cai, X.F., (2005). The theory of city hinterland and the research on the districting ways of hinterland [J]. Economic Geography, 5, p.008.

Chow, R. (1993), 'Things, Common/Places, Passages of the Port City: On Hong Kong and Hong Kong Author Leung Ping-kwan', Differences 5 (3): 179-204.

Chung, C.J. et al. (2001), Great Leap Forward. Cambridge, Massachusetts: Harvard Design School.

Conklin, J. (2001). Wicked problems and social complexity. CogNexus Institute.

Coutard, O. (2008). Placing splintering urbanism: Introduction. Geoforum, Vol. 39, pp.1815-1820.

Cupers, K., (2016). Géographie Volontaire and the Territorial Logic of Architecture. Architectural Histories, 4(1).

[D]:

Davis, D.S. (2000), 'The Consumer Revolution in Urban China', Studies on China 22. Berkeley: University of California Press.

Day, A., (2008). The end of the peasant? New rural reconstruction in China. Boundary 2, 35(2), pp.49-73.

Deal, S., (2017), The Urban Transect: One of the Planning Profession's Most Powerful Tools [Online], available from: http://masglp.olemiss.edu/waterlog/pdf/feb17/wl37.1_article4. pdf, (accessed 25.03.18).

Démurger, S., Fournier, M. and Yang, W., (2010). Rural households' decisions towards income diversification: Evidence from a township in northern China. China Economic Review, 21, pp.S32-S44.

Démurger, S., Sachs, J.D., Woo, W.T., Bao, S., Chang, G. and Mellinger, A., (2002). Geography, economic policy, and regional development in China. Asian Economic Papers, 1(1), pp.146-197.

Dovey, K. (2012). Informal urbanism and complex adaptive assemblage. International Development Planning Review, 34(4), 349-368.

Dreyer, J. (2010), 'The Metropolis and the Capital: Shanghai and Beijing as Paradigms of Space', International Institute for Asian Studies Newsletter 55: 20-21.

Duany, A., Sorlien, S. & Wright, W. (2005). SmartCode Manual: Includes the Complete SmartCode Version 7.5. Miami: Duany Plater-Zyberk & Company.

Duany, A., Plater-Zyberk, E., & Speck, J. (2001). Suburban nation: The rise of sprawl and the decline of the American dream. Macmillan.

Dunham-Jones, E., & Williamson, J. (2011). Retrofitting Suburbia, Updated Edition: Urban Design Solutions for Rede-signing Suburbs. John Wiley & Sons.

[E]:

Eggener, K.L. (2007), 'Placing Resistance: A Critique of Critical Regionalism' in Architectural Regionalism: Collected Writings on Place, Identity, Modernity, and

Tradition, edited by Vincent B. Canizaro, 395-407. New York: Princeton Architectural Press.

Eijkelboom, H. (2007), Paris – New York – Shanghai: A Book about the Past, Present, and (Possibly) Future Capital of the World. New York: Aperture.

Eliasson, K., Johansson, M. and Westlund, H., (2009), February. Labour market aspects of the "new rurality": employ-ment, commuting and entrepreneurship in rural Sweden. In 48th annual meeting of the Western Regional Science Association, Napa Valley, California, February 22e25.

Ellis, E.C., Klein Goldewijk, K., Siebert, S., Lightman, D. and Ramankutty, N., (2010). Anthropogenic transformation of the biomes, 1700 to 2000. Global Ecology and Biogeography, 19(5), pp.589-606.

Ellis, E.C. and Ramankutty, N., (2008). Putting people in the map: anthropogenic biomes of the world. Frontiers in Ecology and the Environment, 6(8), pp.439-447.

Ellis, F., (2000). A framework for livelihoods analysis. In: Rural Livelihoods and Diversity in Developing Countries. Oxford University Press, pp. 28-51.

Endlicher, W., Langner, M., Hesse, M., Mieg, H.A., Kowarik, I., Hostert, P., Kulke, E., Nützmann, G., Schulz, M., van der Meer, E. and Wessolek, G., (2007). Urban ecology– Definitions and concepts. In Shrinking Cities: Effects on Urban Ecol-ogy and Challenges for Urban Development (Vol. 1, No. 16, pp. 1-16). Peter Lang Publishing Group in association with GSE Research.

Eng, I., (1997). The rise of manufacturing towns: externally driven industrialization and urban development in the Pearl River Delta of China. International Journal of Urban and Regional Research, 21(4), pp.554-568.

[F]:

Fabian, L. (2012). Extreme cities and isotropic territories: Scenarios and projects arising from the environmental emer-gency of the central Veneto città diffusa. International Journal of Disaster Risk Science, 3(1), 11-22.

Fan, Jiayang, (2018), "How E-Commerce Is Transforming Rural China. JD.com is expanding its consumer base with drone delivery and local recruits who can exploit villages' tight-knit social networks to drum up business.", The New Yorker, July 23, 2018 Issue.

Fillingham, L. A. (1993).The Order of Things, from Foucault for Beginners, NewYork and London: Writers and Readers Publishing Inc., pp. 78-110.

Fong, V.L. (2004), Only Hope: Coming of Age under China's One-Child Policy. Stanford: Stanford University Press.

Forman, RTT., Godron, M.(1986) Landscape Ecology, Wiley.

Foucault, Michel, (1984). "Of Other Spaces, Heterotopias." Architecture, Mouvement, Continuité 5 (1984): 46-49.

Frankfurt, H.G., 2014. Freiheit und Selbstbestimmung: ausgewählte Texte (Vol. 3). Walter de Gruyter GmbH & Co KG..

[G]:

Geddes, P., (1949). Cities in evolution. William and Norgate Limited, London.

George, L., C.S. (2007). Reproducing Spaces of Chinese Urbanization: New City- Based and Land-Centered Urban Transformation, Urban Studies, Vol. 44, No 9, pp.1827-1855.

Glaeser, E. (2011). Triumph of the city: How our greatest invention makes US richer, smarter, greener, healthier and happier. Pan Macmillan.

Graedel, T.E., 1996. On the concept of industrial ecology. Annual Review of Energy and the Environment, 21(1), pp.69-98.

Graham, S., & Simon Marvin, S. (2001). Splintering Urbanism: Network Infrastructures, Technological Mobilities and the Urban Condition. Routledge,London. pp. 138-146, 160-177.

Greco, C. (2008), Beijing: The New City. Milan: Skira Books.

[H]:

Hamnett, C. (1994), 'Social Polarisation in Global Cities: Theory and Evidence', Urban Studies 31, no. 3: 401-24.

Hamnett, C. and Yulong Shi (2002), 'The Potential and Prospect for Global Cities in China: In the Context of the World System', Geoforum 33, no. 1: 121-35.

Harms, E., (1939). Rural Attitudes in Modern Urban Life. Social Forces, 17(4), pp.486-489.

Harrison, T. (2000). Urban policy: addressing wicked problems. What works, 207-228.

Harvey, D. (2008), 'The Right to the City', New Left Review 253: 23-40.

Harvey, D. (1973), Social Justice and the City. London: Edward Arnold.

Harvey, D. (1972), Society, the City, and the Space-Economy of Urbanism. Washington: Association of American Geo-graphers.

Hill, K., (2009). Urban design and urban water ecosystems. In The water environment of cities (pp. 141-170). Springer, Boston, MA.

Ho, P. (2001). Who owns China's land? Policies, property rights and deliberate institutional ambiguity. The China Quarterly, 394.

Hong, Y.-H. (2007).Assembling land for urban development: Issues and Opportunities. In: Hong, Y.-H. & Needham, B. (eds.) Analyzing land readjustment : economics, law, and collective action. Cambridge, Mass. : Lincoln Institute of Land Policy.

Hou, X., (2013). Community capitalism in China: The state, the market, and collectivism. Cambridge University Press.

Hou, X., (2011). From Mao to the market: Community capitalism in rural China. Theory, Culture & Society, 28(2), pp.46-68.

Hsing, Y.T., (2010). The great urban transformation: politics of land and property in China. OUP Catalogue.

Huang, Q. (2007), 'The External Trade of Guangdong in Two Thousand Years: Guangzhou as the Center', Shenzhen University Journals 24 (2): 150-157. Shenzhen: Humanity Social Science Press.

Huang, Y. (2008), Capitalism with Chinese Characteristics: Entrepreneurship and the State. Cambridge, United King-dom: Cambridge University Press.

Hung, W. (2005), Remaking Beijing: Tiananmen Square and the Creation of a Political Space. London: Reaktion Books.

[I]:

Indovina, F., Matassoni, F.R.A.N.C.A. and Savino, M., (1990). La città diffusa, Venezia, Italy: Daest,pp. 19-45.

Indovina, F. (1998). Algunes consideracions sobre la «ciutat difusa». In Documents d'analisi geografica, pp. 021-32.

Ishiguro, K. (2000), When We Were Orphans. London: Faber & Faber.

[J]:

Jacobs, J. (1961). The death and life of great American cities. Vintage.

Jacoby, S., Lee, C. (2007).Typological formations: renewable building types and the city. Architectural Association.

Jensen, O.B., (2009). Flows of meaning, cultures of movements–urban mobility as meaningful everyday life practice. Mobilities, 4(1), pp.139-158.

Jessop, B., Brenner, N., & Jones, M. (2008). Theorizing sociospatial relations. Environment and Planning. D, Society and space, 26(3), 389.

Jie, W., Zan, Y. and Chun-feng, L., (2005). Two New Methods for Partitioning Port Hinterland [J]. Navigation of China, 3, pp.57-61.

[K]:

Kasimis, C. and Papadopoulos, A.G., (2013). Rural transformations and family farming in contemporary Greece. Re-search in Rural Sociology and Development, 19, pp.263-293.

Knierbein, S. and Viderman, T. eds., 2018. Public Space Unbound: Urban Emancipation and the Post-Political Condi-tion. Routledge.

Kovach, I. and Kristof, L., (2009). The Role of Intermediate Actors in Transmitting Rural Goods and Services in Rural Areas Under Urban Pressure. Journal of Environmental Policy & Planning, 11(1), pp.45-60.

Krause, M., (2013). The ruralization of the world. Public Culture, 25(2 70), pp.233-248.

[L]:

Lane, R., & Woodman, G. (2000). Wicked Problems, Righteous Solutions–Back to the Future on Large Complex Projects. In International Group for Lean Construction Eighth Annual Conference (IGEC), Brighton, England, July.

Lange, O., (1936). On the economic theory of socialism: part one. The review of economic studies, 4(1), pp.53-71

Laurans, V. (2005), 'Shanghai: Modern Conveniences as an Argument for Displacing Residents', China Perspectives, vol. 58: 10-21.

Law, L. (2002), 'Defying Disappearance: Cosmopolitan Public Spaces in Hong Kong', Urban Studies 39 (6): 1625-45.

Lee, C., & Jacoby, S. (2011). Typological urbanism and the idea of the city. Architectural Design, 81(1), 14-23.

Lee, Y. (1992), 'Rural Transformation and Decentralized Urban Growth in China' in Urbanizing China, edited by Gregory E. Guldin. New York: Greenwood Press.

Lefebvre, H. (1991), The Production of Space (Nicholson-Smith, D. trans.). Cambridge, United Kingdom: Blackwell.

Lefebvre, H., (1996). The right to the city. Writings on cities, pp.63-181.

Lefebvre, H. (2003). The urban revolution. U of Minnesota Press.

Leong, C.M.L., Pan, S.L., Newell, S. and Cui, L., (2016). The Emergence of Self-Organizing E-Commerce Ecosystems in Remote Villages of China: A Tale of Digital Empowerment for Rural Development. Mis Quarterly, 40(2), pp.475-484.

Li, B.,(2015), China's hukou reform a small step in the right direction,[Online], available from: http://www.eastasiaforum.org/2015/01/13/chinas-hukou-reform-a-small-step-in-theright-direction/, (accessed 16.11.15).

Li, L.-H. & Li, X. (2007). Land Readjustment: An Innovative Urban Experiment in China.Urban Studies, 44, 81.

Li, L. H. (2006). Public-private partnership - the solution of the property rights problems in China's urban villages? A case study of Shenzhen. The first Fudan International Urban Forum. Fudan University, Shanghai, China.

Li, W. (ed. 2006). Introduction, from Urban Enclave to Ethnic Suburb, Honolulu: University of Hawai'i Press, pp. 1-17.

Lin, G. C. S. (2007). Reproducing spaces of Chinese urbanization: new city-based and land-centred urban transforma-tion. Urban Studies, 44(9), 1827-1855.

Lin, H (2014), Making Housing Affordable in Fast-Growing Chinese Cities: A Shenzhen Perspective,[Online], available from: http://www.newtowninstitute.org/spip.php?article1064, (accessed 19.11.15).

Lin, J.Y. and Chen, B., (2011). Urbanization and urban-rural inequality in china: a new perspective from the govern-ment's development strategy. Frontiers of Economics in China, 6(1), pp.1-21.

Liu, S., Li, X. and Zhang, M., (2003). Scenario analysis on urbanization and rural-urban migration in China. International Institute for Applied Systems Analysis, Vienna.

Liu, Y., He, S., Wu, F., & Webster, C. (2010). Urban villages under China's rapid urbanization: unregulated assets and transitional neighbourhoods. Habitat International, 34(2),135-144.

Liu, Z. (2005). Institution and inequality: the hukou system in China. Journal of comparative economics, 33(1), 133-157.

Logan, J.R. (2002), The New Chinese City: Globalization and Market Reform. Oxford: Blackwell.

Lu, K.R. and Jin, J., (2015). From Lost Villager Autonomy to Diversified Collaborative Governance——A Case Study on the Innovative Practices of Rural Governance in Jinhua

City of Zhejiang Province. Journal of Guangxi University for Nationalities (philosophy and Social Science Edition), 3, p.011.

Lu, M. and Wan, G., (2014). Urbanization and urban systems in the People's Republic of China: research findings and policy recommendations. Journal of Economic Surveys, 28(4), pp.671-685.

Lulu,F. (2019),Taobao Villages - The Emergence of a New Pattern of Rural Ecommerce in China and its Social Implica-tions, Retrieved 16.09.2019, https://www.fes-indonesia.org/e/en-taobao-villages-the-emergence-of-a-new-pattern-of-rural-ecommerce-in-china-and-its-social-implications/

Luo Q., and Tao W. (2004), 'Research about the Relationship of Exhibitions and Urban Economical Social Development: Guangzhou Fair as example', Journal of Beijing International Studies University, no. 3: 30-37.

[M]:

Makinen, J. (2016), 'Chinese e-commerce giant Alibaba connects rural residents to online shopping', The L.A. Times, [Online], available from: https://www.latimes.com/world/asia/la-fg-china-rural-economy-20160403-story.html, (ac-cessed 27.11.19).

Marcuse, P. (2006), 'Space in the Globalizing City' in The Global Cities Reader, edited by Neil Brenner and Roger Keil, 361-9. London: Routledge.

Marshall, R. (2003), Emerging Urbanity: Global urban projects in the Asia Pacific Rim. London: Spon Press.

Marton, A.M., (1999). Rural industrialization in China's lower Yangzi delta: Institutionalizing transactional networks. GeoJournal, 49(3), pp.245-255.

Mc Gee, T. (2016) Ćitta Diffusà and "Desakotasasi." Comparing "Diffuse Urbanization" in Europe and Asia: Some Pre-liminary Thoughts on a "Project in the making". Working seminar EFPL-IUAV PhD Masterclass Horizontal Metropolis held in Venice-Certosa, August 23-26, 2016.

McGee, T. G. (2010). Building liveable cities in Asia in the twenty–first century research and policy challenges for the urban future of Asia. Malay J Environ Manag, 11(1), 14-28.

McGee, T., (2009). UNU-IAS Working Paper No. 161.

McGee, T. G. (2008). Managing the rural–urban transformation in East Asia in the 21st century. Sustainability Science, 3(1), 155-167.

McGee, T. G. (2008). The Emergence of Desakota Regions in Asia: Expanding a Hypothesis, from Norton Ginsburg, Bruce Koppel, and Terry McGee (eds.) Theextended Metropolis: Settlement Transition in Asia, Honolulu: University of HawaiiPress.

McGee, T.G. (2007), China's Urban Space: Development under Market Socialism. London: Routledge.

Meng, L., (2012). Can grain subsidies impede rural–urban migration in hinterland China? Evidence from field surveys. China Economic Review, 23(3), pp.729-741.

Ming, Z.Y.Q., (2003). Rurbanization and Strategical Choice of Urbanization in Southern Jiangsu Province. Academic Journal of Suzhou University, 4, p.021.

Müller, S., Rudolph, C. and Janke, C., (2018). Drones for future mobility: Baloney or part of the solution?.

Murgante, B. and Danese, M., (2012). Urban Versus Rural: The Decrease of Agricultural Areas. New Technologies for Constructing Complex Agricultural and Environmental Systems, p.154.

Murphy, R. (2002), How Migrant Labor Is Changing Rural China. Cambridge, United Kingdom: Cambridge University Press.

[N]:

Naughton, B. (2007). The Chinese Economy: Transitions and Growth. Cambridge, Massachusetts: Massachusetts Insti-tute of Technology Press.

Neuman, M. and Zonneveld, W., (2018). The resurgence of regional design. European Planning Studies, 26(7), pp.1297-1311.

Neuman, M., (2000). Regional design: Recovering a great landscape architecture and urban planning tradition. Land-scape and Urban Planning, 47(3-4), pp.115-128.

Ng, M. K. (2002). Sustainable urban development issues in Chinese transitional cities: Hong Kong and Shenzhen. In-ternational Planning Studies, 7, 7.

[P]:

Pallarès-Blanch, M., (2009). The benefits of Nature Reserve Areas in local development: An opportunity to develop a sustainable strategy in peripheral areas. Naturbanization: New identities and processes for rural-natural areas.

Pan, J. and Liu, W. (2014). "Quantitative Delimitation of Urban Influential Hinterland in China." Journal of Urban Plan-ning and Development, (Online publication date: 1-Dec-2015).

Pan, J., Shi, P. and Dong, X., (2008). Measurements for Urban Hinterland Area of Cities at Prefecture Level or above in China [J]. Acta Geographica Sinica, 6, p.010.

Preston, F., (2012). A global redesign? shaping the circular economy. Energy, Environment and Resource Governance, 2, pp.1-20.

Purcell, M. (2002). Excavating Lefebvre: The right to the city and its urban politics of the inhabitant. GeoJournal, 58(2), 99-108.

[Q]:

Qin, H.Z. and Yang, T., (2014). The Culture Practice in the Rurbanization——A Case Study on Wutai New Village in Central Guangxi. Journal of Guangxi University for Nationalities (Philosophy and Social Science Edition), 5, p.008.

Quertamp, F. and de Miras, C., (2012). Periurbanization and governance of large metropolises in Vietnam. TRENDS OF URBANIZATION AND SUBURBANIZATION IN SOUTHEAST ASIA, 75.

[R]:

Reghezza-Zitt, M., Rufat, S., Djament-Tran, G., Le Blanc, A. and Lhomme, S., (2012). What resilience is not: uses and abuses. Cybergeo: European Journal of Geography.

Ren, X. (2014), Urban Design and Inequality: The Case of Urban Villages in China's Megacities,[Online], available from:http://citiespapers.ssrc.org/urban-design-and-inequality-the-case-of-urban-villages-in-chinas-megacities/,(accessed 19.11.15).

Ren, X. (2011), Building Globalization: Transnational Architecture Production in Urban China.Chicago: University of Chicago Press.

Rittel, H. W., & Webber, M. M. (1973). Dilemmas in a general theory of planning. Policy sciences, 4(2), 155-169.

Rittel, H. W. (1972). On the Planning Crisis: Systems Analysis of the" First and Second Generations" (pp. 390-396). Institute of Urban and Regional Development.

Rode, P., Burdett, R., Brown, R., Ramos, F., Kitazawa, K., Paccoud, A., & Tesfay, N. (2009). Cities and social equity: in-equality, territory and urban form. Urban Age Programme, London School of Economics.

Rodin, J., (2014). The resilience dividend: being strong in a world where things go wrong. PublicAffairs

Rowe, P.G. (2005), East Asia Modern: Shaping the Contemporary City. London: Reaktion.

Rowe, P.G. (2002), Architectural Encounters with Essence and Form in Modern China. Cambridge, Massachusetts: Massachusetts Institute of Technology Press.

Roy, A. (2009). The 21st-century metropolis: new geographies of theory. Regional Studies, 43(6), 819-830.

[S]:

Sassen, S. (1991), The Global City. Princeton: Princeton University Press.

Sassen, S. (2000). Spatialities and Temporalities of the Global: Elements for a Theorization. Public Culture, 12(1), 215-232.

Sassen, S. (2000). Territory and territoriality in the global economy. International Sociology, 15(2), 372-393.

Sassen, S. (2002). Locating cities on global circuits. Environment and urbanization, 14(1), 13-30.

Sassen, S., (2013). Expelled: Humans in Capitalism's Deepening Crisis. Journal of World-Systems Research, 19(2), p.198.

Satterthwaite, D., (2011), Upgrading Dense Informal Settlements ,[Online], available from: https://lsecities.net/media/objects/articles/upgrading-dense-informal-settlements/en-gb/,(accessed 16.11.15).

SCMP,(2013),China's left-behind kids cloud the country's dream, [Online], available from: http://www.scmp.com/topics/hukou-system,(accessed 16.11.15).

Scott, A.J., Carter, C., Reed, M.R., Larkham, P., Adams, D., Morton, N., Waters, R., Collier, D., Crean, C., Curzon, R. and Forster, R., (2013). Disintegrated development at the rural–urban fringe: Re-connecting spatial planning theory and practice. Progress in planning, 83, pp.1-52.

Secchi, B., & Ingallina, P. (2006). Première leçon d'urbanisme. Parenthèses.

Secchi, B. (2007). Section 1: Wasted and Reclaimed Landscapes-Rethinking and Redesigning the Urban Landscape. Places, 19(1).

Secchi, B., & Viganò, P. (2009). Antwerp-Territory of a new modernity, Explorations in/of urbanism.

Secchi, B. (2010). A new urban question. Territorio.

Secchi, B., & Viganò, P. (2011). La ville poreuse: un projet pour le Grand Paris et la métropole de l'après-Kyoto. MētisPresses.

Secchi B., Vigano P. (2013), Habiter le Grand Paris, l'habitabilité des territoires : cycles de vie, continuité urbaine, métropole horizontale, , [Online], available from: http://www.ateliergrandpa-ris.fr/aigp/conseil/studio/Studio13Habiter2013.pdf, (accessed 25.08.16).

Seto, K. C., & Shepherd, J. M. (2009). Global urban land-use trends and climate impacts. Current Opinion in Environ-mental Sustainability, 1(1), 89-95.

Shane, D. G. (2017) Professor in the Urban Design program at Columbia GSAPP. (Personal communication, 29th Au-gust 2017).

Shane, D. G. (2019) Email communication sent to Dan Narita, 14th March.

Silk, R.,(2014), China's Hukou Reform Plan Starts to Take Shape ,[Online], available from: http://blogs.wsj.com/chinarealtime/2014/08/04/chinas-hukou-reform-plan-starts-to-takeshape/, (accessed 16.11.15).

Sklair, L. (2009), 'The emancipatory potential of generic globalization', Globalizations 6/4: 523-537.

Smart, A., & Smart, J. (2001). Local citizenship: welfare reform urban/rural status, and exclusion in China. Environment and Planning A, 33(10), 1853-1870.

Smith, R.G. (2003), 'World City Topologies', Progress in Human Geography 27 (5): 561-82.

Soja,E. (2000) Putting Cities First, from Postmetropolis. Oxford, UK andMalden, MA: Blackwell, pp. 19-49.

Soja, E.W. (1996). Thirdspace: Journeys to Los Angeles and other real-and-imagined places (p. 53). Oxford: Blackwell.

Song-di, W.U., (2004). Trade Ports and their Hinterland and The space course of China's Modernization [J]. Hebei Academic Journal, 3, p.028.

Soureli, K., & Youn, E. (2009). Urban Restructuring and the Crisis: A Symposium with Neil Brenner, John Friedmann, Margit Mayer, Allen J. Scott, and Edward W. Soja. Critical Planning, 16, 35-58.

Spence, M., Annez, P.C. and Buckley, R.M. eds., (2008). Urbanization and growth. World Bank Publications.

Strickland R ed., (2005). Post-Urbanism & ReUrbanism (Michigan Debates on Urbanism), Taubman College of Archi-tecture and Urban Planning.

[T]:

Tan, M., Li, X., Xie, H. and Lu, C., (2005). Urban land expansion and arable land loss in China—a case study of Beijing–Tianjin–Hebei region. Land use policy, 22(3), pp.187-196.

Tang, W.S. and Chung, H., (2002). Rural–urban transition in China: illegal land use and construction. Asia Pacific View-point, 43(1), pp.43-62.

Tao, R., & Xu, Z. (2007). Urbanization, rural land system and social security for migrants in China. The Journal of De-velopment Studies, 43(7), 1301-1320.

Taylor, J.R., (2015). The China dream is an urban dream: Assessing the CPC's national new-type urbanization plan. Journal of Chinese Political Science, 20(2), pp.107-120.

Tombesi, P., Dave, B., and Scriver, P. (2003), 'Routine production or symbolic analysis? India and the globalization of architectural services', Journal of Architecture 8: 63-94.

Tong, D. & Feng, C. (2009). Review and Progress of the research on Urban Villages. Human Geography (Chinese), 24.

[U]:

Umweltbundesamt.(2013), MUFLAN – Multifunktionale Landschaften, [Online], available from: http://www.umweltbundesamt.at/fileadmin/site/publikationen/REP0419.pdf, (accessed 30.08.16).

[V]:

Van Bueren, E. M., Klijn, E. H., & Koppenjan, J. F. (2003). Dealing with wicked problems in networks: Analyzing an envi-ronmental debate from a network perspective. Journal of Public Administration Research and Theory, 13(2), 193-212.

Vasilevska, L., (2010). Rural development and regional policy: Conceptual framework. Facta universitatis-series: Archi-tecture and Civil Engineering, 8(3), pp.353-359.

Vejre, H., Abildtrup, J., Andersen, E., Andersen, P.S., Brandt, J., Busck, A., Dalgaard, T., Hasler, B., Huusom, H., Kristensen, L.S. and Kristensen, S.P., (2007). Multifunctional agriculture and multifunctional landscapes - land use as an interface. In Multifunctional land use (pp. 93-104). Springer Berlin Heidelberg.

Viganò, P. (1999). La città elementare, Milan, Skira.

Viganò, P. (2008). Water and Asphalt: The Project of Isotropy in the Metropolitan Region of Venice. Architectural De-sign, 78(1), 34-39.

Viganò, P. (2010). EXTREME CITY: A DESIGN AND A RESEARCH THEME. EXTREME CITY, climate change and the trans-formation of the waterscape, 11.

Viganò, P. (2012). Extreme cities and bad places. International Journal of Disaster Risk Science, 3(1), 3-10.

Viganò, P. (2013). Urbanism and Ecological Rationality. In Resilience in Ecology and Urban Design (pp. 407-426). Springer Netherlands.

Viganò, P., (2015). State of crisis and project: the Horizontal Metropolis (No. EPFL-CHAPTER-215439). Jovis.

Vigano, P. (2015) "The Horizontal Metropolis": a radical project. In: Barcelloni, M. and Cavalieri, C. (eds.) Lab-U, EPFL 2015: VIII International PhD Seminar 'Urbanism & Urbanization' Symposium Latsis EPFL 2015, 12–14 October 2015, Switzerland. Lausanne, EPFL. pp. 6-7.

Viganò, P., Cavalieri, C. and Barcelloni Corte, M., 2016. The Horizontal Metropolis: a radical project. Anthos Magazine, 2(EPFL-ARTICLE-215432).

Viganò, P.(2016),The Territories of Urbanism: The project as knowledge producer. EPFL Press.

Vlassenrood, L.(2015), Chinese urbanization through the lens of Da Lang, [Online], available from: http://www.newtowninstitute.org/spip.php?article1074,(accessed 19.11.15).

Von Bertalanffy, L., (1950). An outline of general system theory. British Journal for the Philosophy of science.

[W]:

Wang, F. L. (2010), 'Conflicts, Resistance, and the Transformation of the Hukou System' in Chinese Society: Change, Conflict and Resistance, edited by Elizabeth J. Perry and Mark Selden, 80-100. London: Routledge.

Wang, H. (2003), 'The New Criticism' in One China, Many Paths, edited by Chaohua Wang,55-86. London: Verso.

Wang, J. and Wang, X., (2015). New urbanization: A new vision of China's urban–rural development and planning. Frontiers of Architectural Research, 4(2), pp.166-168.

Wang, M. Y. (2002). Small city, big solution? China's hukou system reform and its potential impacts. disP-The Planning Review, 38(151), 23-29.

Wang, Z., Deng, X. and Wong, C., (2016). Integrated Land Governance for Eco-Urbanization. Sustainability, 8(9), p.903.

Whelton, M., & Ballard, G. (2002, August). Wicked problems in project definition. In Proceedings of the 10th annual conference of the International Group for Lean Construction (pp. 6-8).

Windrow, H., & Guha, A. (2005). Hukou System, Migrant Workers, & (and) State power in the People's Republic of China, The. Nw. Univ. J. Int'l Hum. Rts., 3, 1.

Wing Chan, K., & Buckingham, W. (2008). Is China abolishing the hukou system? The China Quarterly, 195, 582-606.

Wong, J.K.Y. (2002), 'The Distance between Hong Kong and China: Hong Kong People's Perception of Their Identity', paper presented at the Crossroads in Cultural Studies: Fourth International Conference, Tampere, Finland.

Woods, M. (2007), 'Engaging the Global Countryside: Globalization, Hybridity and the Reconstitution of Rural Place', Progress in Human Geography 31 (4): 485-507.

Wu, F., Zhang, F., & Webster, C. (2013). Rural migrants in urban China: enclaves and transient urbanism (Vol. 104). Routledge.

Wu, F. (2004), 'Residential Relocation under Market-Oriented Redevelopment: The Process and Outcomes in Urban China', Geoforum 35: 453-70.

[X]:

Xu, D., Deng, X., Guo, S. and Liu, S., (2018). Sensitivity of livelihood strategy to livelihood capital: an empirical investi-gation using nationally representative survey data from rural China. Social Indicators Research, pp.1-19.

Xu, J. and Yeh, A.G.O. (eds.) (2007), Urban Development in Post-Reform China: State, Market, and Space. London: Routledge.

[Y]:

Yang, Y., (2000). The System of Farmland in China: An Analytical Framework [J]. Social sciences in China, 2, p.005.

Yansui, L., (2007). Rural transformation development and new countryside construction in eastern coastal area of China. Acta Geographica Sinica, 62(6), pp.563-570.

Yeung, Y. M., & Shen, J. (2008). The Pan-Pearl River Delta: An Emerging Regional Economy in a Globalizing China. Chinese University Press.

Young, R.F., (2017). "Free cities and regions"—Patrick Geddes's theory of planning. Landscape and urban planning, 166, pp.27-36.

Yu, W. and Zhou, W., 2017. The spatiotemporal pattern of urban expansion in China: A comparison study of three urban mega regions. Remote Sensing, 9(1), p.45.

[Z]:

Zhan, J. and Lu, Q., (2003). Assessment and Empirical Study on the Relationship between Infrastructure Construction and Urban-Rural Development [J]. Acta Geographica Sinica, 4, p.017.

Zhang, H., Ma, W.C. and Wang, X.R., (2008). Rapid urbanization and implications for flood risk management in hinter-land of the Pearl River Delta, China: The Foshan study. Sensors, 8(4), pp.2223-2239.

Zhang, L. (2001), 'Migration and Privatization of Space and Power in Late Socialist China', American Ethnologist 28, no. 1: 179-205.

Zhang, L.M. and Ong, A. (eds.) (2008), Privatizing China: Socialism from Afar. Ithaca: Cornell University Press.

Zhan, S. and Andreas, J. (2015), June. Beyond the countryside: Hukou reform and agrarian capitalism in China. In Con-ference on Land grabbing, Conflict, and Agrarian-Environmental Transformations: Perspectives from East and South-east Asia, Chiang Mai University (Vol. 56).

Zhang, X. (2009), 'Competitiveness and International Growth: The External Trade and Construction Development of Guangzhou (1912-1936)', The Research of Chinese Social Economy History (4): 91-102.

Zhang, X. (2008), Postsocialism and Cultural Politics: China in the Last Decade of the Twentieth Century. Durham: Duke University Press.

Zhu, Y. (2007). China's floating population and their settlement intention in the cities: Beyond the Hukou reform. Habitat International, 31(1), 65-76.

www.ingramcontent.com/pod-product-compliance
Ingram Content Group UK Ltd.
Pitfield, Milton Keynes, MK11 3LW, UK
UKHW022001190726
13853UKWH00004B/1672

9 783736 975170